FLORA ZAMBESIACA

Flora terrarum Zambesii aquis conjunctarum

VOLUME NINE: PART SIX

FLORA ZAMBESIACA

MOZAMBIQUE

MALAWI, ZAMBIA, ZIMBABWE

BOTSWANA, CAPRIVI STRIP

VOLUME NINE: PART SIX

Edited by
E. LAUNERT & G. V. POPE

on behalf of the Editorial Board:

G. Ll. LUCAS
Royal Botanic Gardens, Kew

E. LAUNERT
Natural History Museum

I. MOREIRA
Centro de Botânica, Instituto de Investigação
Cientifíca Tropical, Lisboa

G. V. POPE
Royal Botanic Gardens, Kew

Published by the Managing Committee on behalf of
the contributors to Flora Zambesiaca
1991

Typeset at the Royal Botanic Gardens, Kew, by
Pam Arnold, Christine Beard, Brenda Carey,
Margaret Newman, Pam Rosen
and Helen Ward

Printed in Great Britain by
Whitstable Litho Printers Ltd.,
Whitstable, Kent

ISBN 0 565 01145 6

CONTENTS

LIST OF FAMILIES INCLUDED IN
VOLUME IX, PART 6

Due to circumstances beyond our control we have not been able to include family number 158 (Myricaceae) in this part. This will now appear in a later part.

154. ULMACEAE

By C.M. Wilmot-Dear

Trees or shrubs, monoecious or dioecious, sometimes spiny. Leaves alternate, simple; lamina often unequal-sided at base; stipules lateral, caducous. Flowers small, unisexual or bisexual, regular, axillary, solitary or in cymes or clusters. Sepals 4–5(8), imbricate or valvate, free or shortly united, persistent. Petals absent. Stamens as many as, and opposite to, the calyx lobes or (not in south tropical Africa) a few more, inserted at the base of the calyx, erect in bud; anthers 2-thecous, opening longitudinally. Ovary superior, of 2 united carpels, 1(2)-locular; styles 2, divergent; ovule solitary, pendulous from or near apex, anatropous. Fruit thinly fleshy or compressed, dry and ± winged or appendiculate; endocarp hard. Seeds without endosperm; embryo curved or (not in south tropical Africa) straight.

A family of about 14 genera, with some 120 species, mainly tropical and north temperate in distribution. Four genera are represented in Africa, three of which occur in the Flora Zambesiaca area.

Ulmus parvifolia is sometimes cultivated as an ornamental. A brief description of this species is as follows: Deciduous trees with fissured bark. Leaves serrate or biserrate. Flowers bisexual or hermaphrodite. Perianth herbaceous, cup-shaped, with 4–8 connate lobes. Stamens equal in number to the perianth lobes. Ovary compressed, stipitate. Fruit a samara with a wing surrounding the fruit, emarginate at the apex.

The specimens *Balsinhas* 1799 (COI; LISC) and *Barbosa & Lemos* 8264 (COI; LISC) of *Ulmus parvifolia* were taken from plants cultivated in the Jardim Tunduru (Vasco da Gama), Maputo and in Namaacha respectively.

1. Plant with axillary spines; stipules united along one margin - - - **1. Chaetacme**
– Plant unarmed (in African species); stipules free - - - - - - - 2
2. Ovary stipitate, compressed; fruit a flat samara with the wing surrounding the fruit; embryo straight. (Cultivated). - - - - - - - - - - - **Ulmus**
– Ovary sessile; fruit a thinly fleshy drupe; embryo curved - - - - - 3
3. Male flowers with induplicate-valvate calyx lobes; leaves serrate almost from the base; styles 0.5–1(2) mm. long - - - - - - - - - - - **2. Trema**
– Male flowers with imbricate calyx lobes; leaves entire, coarsely toothed or serrate in the upper two-thirds; styles 2–5 mm. long - - - - - - - - **3. Celtis**

1. CHAETACME Planch.

Chaetacme Planch. in Ann. Sci. Nat., Bot., sér. 3, **10**: 266; 340 (1848).

Trees or shrubs, monoecious or dioecious; branches with axillary spines. Leaves penninerved, shortly petiolate, apex long-mucronate, base slightly unequal-sided; stipules large, united along one margin, enclosing the terminal bud, caducous. Inflorescences cymose, often branched, usually congested, entirely male or with 1(2) females near the base; remaining female flowers solitary and usually in the upper axils. Sepals 5, shortly basally united, with the male buds induplicate-valvate, and the female buds imbricate. Female flowers without staminodes; male flowers with pistillode, stamens equal in number to sepals. Ovary sessile, unilocular; styles 2, long, unbranched, divaricate, persistent. Fruits large, endocarp very hard. Embryo curved.

A genus of one species, confined to Africa and Madagascar.

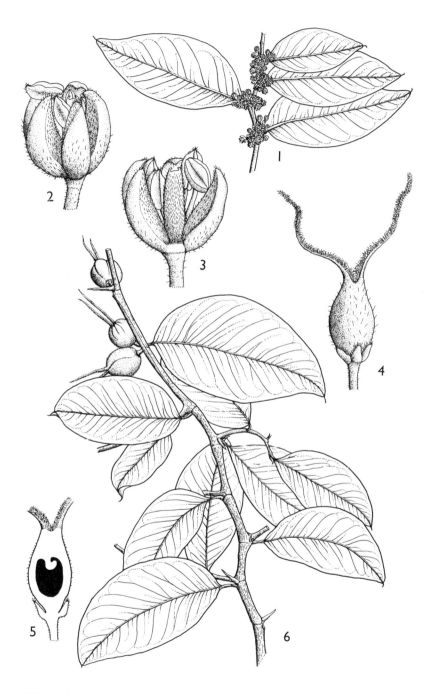

Tab. 1. CHAETACME ARISTATA. 1, male flowering branchlet (×⅔); 2, male flower (× 10); 3, male flower with one sepal removed (× 10), 1–3 *Paulo* 733; 4, female flower (× 6); 5, longitudinal section of female flower (× 10), 4–5 *Bally* 8965; 6, fruiting branchlet (×⅔), *Semsei* 3085. Drawn by M.E. Church. From F.T.E.A.

Chaetacme aristata Planch. in Ann. Sci. Nat., Bot., sér. 3, **10**: 341 (1848). —Engl., Pflanzenw. Ost-Afr., **C**: 160 (1895). —Sim, For. Fl. Col. Cape Good Hope: 305, t. 160 (1907). —Engl., Pflanzenw. Afr.: 15, t. 8 (1915). —N.E. Brown in F.C. **5**, 2: 521 (1925). —Burtt Davy, Fl. Pl. & Ferns Transv. **2**: 437 (1932). —Brenan, Check-list For. Trees Shrubs Tang. Terr. **5**, 2: 625 (1949). —Andrews, Fl. Pl. Anglo-Egypt. Sudan **2**: 254, t. 88 (1952). —Keay in F.W.T.A. ed. 2, **1**: 593 (1958). —F. White, F.F.N.R.: 22 (1962). —Polhill in Kew Bull. **19**: 144 (1964). —von Breitenbach, Indig. Trees Southern Afr. **2**: 84 (1965). —Polhill in F.T.E.A., Ulmaceae: 12 (1966). —Palmer & Pitman, Trees of S. Afr. **1**: 433 & photo (1972). —Compton, Fl. Swaziland: 173 (1976). —K. Coates Palgrave, Trees Southern Africa: 98 (1977). TAB. **1**. Type from S. Africa (Cape Province).

Chaetacme nitida Planch. & Harv. in Harv. Thes. Cap. **1**: 16, t. 25 (1859). Types from S. Africa (Cape Province).

Chaetacme madagascariensis Bak. in Journ. Linn. Soc., Bot. **21**: 443 (1885). —Leroy in Fl. Madag., fam. 54: 14 (1952). —Aubreville, Fl. For. Cote d'Ivoire **1**: 45, t. 3, fig. 7–9 (1954). Type from Madagascar.

Chaetacme serrata Engl. in Not. Bot. Gart. Mus. Berl. **3**: 24 (1900). Types from Tanzania and S. Africa (Cape Province).

Chaetacme aristata var. *longifolia* Engl. ex de Wild. & T. Durand in Ann. Mus. Congo, Bot. **3**, Reliq-Dewevr: 214 (1901). Type from Zaire.

Bosqueia spinosa Engl., Bot. Jahrb. **40**: 548 (1908). Type from Tanzania.

Chaetacme aristata var. *nitida* (Planch. & Harv.) Engl., Pflanzenw. Afr.: 15 (1915). Type as for *C. nitida.*

Chaetacme microcarpa Rendle in F.T.A. **6**, 2: 13 (1916). —Hutch., Dalz. & Moss in F.W.T.A. **1**: 423 (1928). —Hauman in F.C.B. **1**: 51 (1948). Type from Sudan.

Shrub tending to scramble, or small spreading much-branched tree, 1–13 m. tall; branches ± drooping, zigzag; bark smooth, grey, later fibrous and longitudinally striate; twigs shortly erect-pubescent to glabrous, often short and becoming spinose at apex; axillary spines (single) paired, 4–30 mm. long. Leaves (2)3–9(11) × (1)1.5–4.5(5) cm., elliptic to elliptic-ovate; apex acute to shortly acuminate and aristate with a bristle up to 7 mm. long; base broadly cuneate to rounded (to slightly cordate), ± asymmetrical; margin entire or serrate with few irregularly-spaced (or many regular, especially in juvenile plants) mucronate teeth; lamina coriaceous (juvenile foliage sometimes membranous-chartaceous), glabrous and very glossy dark green above, paler and often scabrid or sparsely fine-pubescent below, especially on midrib; midrib prominent beneath, ± grooved above, lateral veins c. 10, looping near margin, indistinct; petiole 3–6 mm. long; stipules (5)10–20 mm. long, narrowly oblong-lanceolate, caducous. Inflorescences 5–15 mm. long, flowers greenish-cream, subtended by small ovate ± amplexicaul bracts, c. 1 × 1 mm., pedicels 1–3 mm. long, sepals finely pubescent. Male flowers (few) 10–30 (or more), sepals 2–3.5 × 1.5–2 mm. Female flowers with sepals 1–2 × 1–1.5 mm.; ovary sparsely fine-pubescent, 3–5 mm. long; styles 7–20(30) mm. long, persistent but easily dislodged when dry. Fruit green, ripening yellowish-orange, ovoid or depressed-globose, 10–15 × 10–14 mm. (dried), up to 35 × 20 mm. (fresh), glabrous.

Botswana. SE: Old Palapye, fr. 26.x.1980, *Woollard* 818 (SRGH). **Zambia**. N: L. Kashiba, fl. 22.x.1957, *Fanshawe* 3800 (K). S: Mazabuka Distr., Ngongo R., fr. 13.i.1960, *White* 6233 (K; SRGH). **Zimbabwe**. C: Mt. Wedza, st. 2.i.1936, *McGregor* 29/37 (K; SRGH). E: Mutare, Main street, fl. 9.iii.1962, *Chase* 7648 (COI; K; PRE; SRGH). S: Chipinge Distr., c. 2 km. W. of Musirizwe/Bwazi R. confluence, st. 29.i.1975, *Pope, Biegel & Russell* 1415 (BM; K; SRGH). **Malawi**. S: Mulanje Distr., Litchenya Forest Reserve, near Mabuka Court, st. 6.ix.1970, *Müller* 1558A (K; SRGH). **Mozambique**. Z: Mt. Mututche, fr. 2.v.1972, *Howard* 671 (SRGH). MS: Mt. Zembe, st. 20.vii.1970, *Müller & Gordon* 1359 (SRGH). GI: entre o Acampamento do Lumane e estrada Xai-Xai (Vila Joao Belo) - Macia, fl. 21.ix.1948, *Myre & Carvalho* 263 (K; SRGH). M: Maputo, st. 20.i.1947, *Hornby* 2560 (BM; SRGH).

Also from West Africa to the Sudan, east tropical Africa and Zaire to S. Africa (Transvaal, Natal and Cape Province) and in Madagascar. Low and medium altitude mixed evergreen forest, riverine forest or thickets, coastal forests and sand dunes, stream sides in wooded grassland, often in rocky places; 600–1500 m.

Distinctive in its axillary spines and glossy leaves conspicuously bristle-tipped at the apex.

2. TREMA Lour.

Trema Lour., Fl. Cochinch.: 562 (1790).

Trees or shrubs, monoecious or dioecious, unarmed. Leaves alternate, penninerved, petiolate, unequal-sided at the base, serrate; stipules lateral, paired, free, caducous. Inflorescences axillary, cymose, often branched, usually congested, mainly male with female and bisexual flowers fewer and borne towards the apex. Sepals (4)5, shortly united at the base, induplicate-valvate in male buds, and imbricate in female buds. Ovary sessile, 1-locular; styles short, unbranched, divaricate or inrolled, persistent. Fruits small, endocarp hard.

A genus of about 15 species in the tropics and subtropics, only one present in Africa.

Trema orientalis (L.) Blume, Mus. Bot., Lugd.-Bat. **2**: 62 (1852). —Leroy in Fl. Madag., fam. 54: 10 (1952). —Polhill in Kew Bull. **19**: 143 (1964); in F.T.E.A., Ulmaceae: 10 (1966). —Schreiber in Merxm., Prod. Fl. SW. Afr., fam. 15: 1 (1967). —Palmer & Pitman, Trees of S. Afr. **1**: 429, photo (1972). —Compton, Fl. Swaziland: 173 (1976). —K. Coates Palgrave, Trees Southern Africa: 98 (1977). TAB. **2**. Types from Sri Lanka.

　　Celtis orientalis L., Sp. Pl.: 1044 (1753).
　　Celtis guineensis Schumach. & Thonn., Beskr. Guin. Pl: 160 (1827). Type from Ghana.
　　Sponia bracteolata Hochst. in Flora **28**: 87 (1845). Type from S. Africa (Natal).
　　Sponia glomerata Hochst. in Flora **28**: 87 (1845). Type from S. Africa (Natal).
　　Sponia nitens Hook. & Planch. in Ann. Sci. Nat., Bot. sér. 3, **10**: 325 (1848). Type from Bioko.
　　Trema bracteolata (Hochst.) Blume, Mus. Bot., Lugd.-Bat. **2**: 58 (1852). —Sim, For. Fl. Col. Cape Good Hope: 305, pl. 158 (1907). —N.E. Brown in F.C. **5**, 2: 519 (1925).
　　Trema glomerata (Hochst.) Blume, Mus. Bot., Lugd.-Bat. **2**: 58 (1852).
　　Trema nitens (Hook. & Planch.) Blume, Mus. Bot., Lugd.-Bat. **2**: 58 (1852).
　　Sponia hochstetteri Planch. in DC., Prodr. **17**: 198 (1873). Type from Ethiopia.
　　Trema guineensis (Schumach. & Thonn.) Ficalho, Pl. Ut. Afr. Portug.: 261 (1884). —Engl., Pflanzenw. Ost-Afr. **C**: 160 (1895). —Engl., Pflanzenw. Afr.: 14, fig. 7 (1915). —Rendle in F.T.A. **6**, 2: 11 (1916). —Burtt Davy, Fl. Pl. & Ferns Transv. **2**: 436, t. 67 (1932). —Peter in Fedde, Repert., Beih. **40**, 2: 62 (1932). —Hauman in F.C.B. **1**: 48, t. 8 (1948). —Robyns, Fl. Sperm. Parc Nat. Alb. **1**: 46, t. 4 (1948). —Brenan, Check-list For. Trees Shrubs Tang. Terr. **5**, 2: 625 (1949). —Andrews, Fl. Pl. Anglo-Egypt. Sudan **2**: 256, t. 89 (1952). —Brenan in Mem. N.Y. Bot. Gard. **9**: 76 (1954). —Keay in F.W.T.A. ed. 2, **1**: 592 (1958). —F. White, F.F.N.R.: 24 (1962). —von Breitenbach, Indig. Trees Southern Afr. **2**: 83, t. on page 81 (1965).
　　Trema guineensis var. *hochstetteri* (Planch.) Engl., Pflanzenw. Ost-Afr. **C**: 160 (1895). —Engl., Pflanzenw. Afr.: 14 (1915). —Rendle in F.T.A. **6**, 2: 12 (1916). —Peter in Fedde, Repert., Beih. **40**, 2: 63 (1932).
　　Trema guineensis var. *paucinervia* Hauman in Bull. Jard. Bot. Brux. **16**: 412 (1942); in F.C.B. **1**: 48 (1948). Type from Zaire. See note in Polhill in F.T.E.A., Ulmaceae: 12 (1966).

Shrub or small to medium tree up to 12 m. tall, deciduous; branches ± spreading; bark smooth, grey, lenticels conspicuous; twigs sparsely to densely pubescent. Leaves (4.5)6–11 × (1.5)2.2–5 cm. (juvenile leaves up to 15 × 9 cm.), attenuate-ovate to ovate-lanceolate or oblong-lanceolate; apex acuminate to attenuate; base rounded to cordate, ± asymmetric; margin evenly closely serrate from near the base; lamina ± scabrid, sparsely stiff-hairy (rarely glabrescent) above, ± densely pubescent to tomentose below, ± strongly 3-nerved from the base, midrib and lateral veins grooved above, prominent below; petiole 8–10(13) mm. long; stipules 4–7 mm. long, lanceolate, pubescent, caducous. Cymes 5–10 mm. long, many flowered; flowers mostly male with a few female (bisexual) flowers at the top, flowers greenish-cream; subtending bracts c. 1 mm. long, triangular; pedicels short or absent; sepals 1–1.5(2) × 0.5 mm. Ovary pubescent, 1–1.5 mm. long; styles 0.5–1.2 mm. long, inrolled or divaricate, usually persistent. Fruit drupaceous, purple or black, ovoid-globose, (2)3.5(4) mm. in diam. when dried, glabrous.

　　Zambia. N: Mbesuma, fl. & fr. 10.x.1961, *Astle* 923 (SRGH). W: Mwinilunga Distr., R. Kakema, fl. & fr. 24.viii.1930, *Milne-Redhead* 958 (K; PRE). E: Chipata (Fort Jameson) township, ? planted, fl. 25.iv.1952, *White* 2460 (K). S: Mazabuka Distr., Mochipapa to Sinazongwe, mile 27.6, st. 15.i.1960, *White* 6284 (K; SRGH). **Zimbabwe**. N: Darwin Distr., near upper reaches of Nyarandi R., fl. 27.i.1960, *Phipps* 2438 (K; SRGH). W: Matobo Distr., Mtsheleli Dam, fl. 24.iv.1965, *Crozier* 2165 (SRGH). C: Shurugwi Distr., Wanderer's Valley, c. 10 km. S. of Shurugwi, fl. & fr. 8.xii.1966, *Biegel* 1544 (K; SRGH). E: Mutare, Christmas Pass, fl. & fr. xi.1945, *Chase* 185 (BM; K; SRGH). S: Great Zimbabwe, near Temple, fl. 4.x.1949, *Wild* 3047 (K; PRE; SRGH). **Malawi**. N: Mzimba Distr., Mzuzu, Marymount, fl. & fr. 9.iv.1975, *Pawek* 9191 (K; PRE; SRGH). C: Dedza Distr., Dedza to Mua road, fl. 13.x.1965, *Banda* 689 (SRGH). S: Zomba, Chingale road, fr. 5.i.1956, *Jackson* 1772 (BM; K; SRGH). **Mozambique**. N: Mozambique Distr., Angoche (Antonio Enes), fl. & fr. 16.x.1965, *Mogg* 32247 (PRE; SRGH). T: Estrada Estima-Songo, 11 km. from Estima, fr. 4.ii.1972, *Macêdo* 4769 (SRGH). MS: Kurumadzi R., Jihu, fl. 17.xi.1906, *Swynnerton* 1104 (BM; K; SRGH). GI: Manhiça, near Alvor, fl. 23.iv.1947, *Barbosa* 196 (COI; PRE; SRGH). M: Maputo, encosta da Ponta Vermelha, fl. & fr. 22.i.1960, *de Lemos & Balsinhas* 19 (COI; K; PRE; SRGH).

　　This species occurs throughout Africa south of the Sahara but is absent from west and south Cape Province of S. Africa; also in Madagascar, Mascarene Islands, and tropical Asia. On forest margins, as a pioneer in regenerating forest, in riverine vegetation and in woodland in high rainfall areas at

Tab. 2. TREMA ORIENTALIS. 1, flowering branchlet (× ⅔), *Fundi* 59; 2, male flower (× 16); 3, male flower with one sepal removed (× 16); 4, sepal and stamen (× 20), 2–4 *Vaughan* 2636; 5, female flower (× 20); 6, longitudinal section of female flower (× 20), 5–6 *Fundi* 59; 7, fruit (× 6), *Benedicto* 38. Drawn by M.E. Church. From F.T.E.A.

medium and lower altitudes; also in ravines and valleys, sometimes along dry sandy river beds; 0–2000 m.

3. CELTIS L.

Celtis L., Sp. Pl. **2**: 1043 (1753); Gen. Pl., ed. 5: 467 (1754).

Trees or shrubs, monoecious rarely dioecious, unarmed (often armed in non African species). Leaves petiolate, lamina usually unequal-sided at the base, penninerved or 3-nerved from the base; cystoliths often present, giving a scabrous texture to the mature leaves; stipules usually small. Flowers often precocious, in many small cymes (sometimes fasciculate in non African species), female flowers sometimes in separate inflorescences. Sepals (4)5, ± free, imbricate. Anthers ovate. Ovary sessile; styles often once or twice-branched. Fruit drupaceous, thinly fleshy, sometimes with ribbed endocarp.

A genus of over 50 species, widespread in tropical and temperate regions; fewer than 10 species occur in Africa, with 4 recorded from the Flora Zambesiaca region.

1. Leaves with the basal lateral nerves extending well into the upper half of the blade; main lateral nerves 1–2(3) on each side of midrib; fruit subglobose or ovoid - - - - - 2
– Leaves with the basal lateral nerves not or hardly extending into upper half of the blade; main lateral nerves (2)3–6 on each side of midrib; fruit usually ellipsoid-ovoid or conical-ovoid, 4-ribbed when dry - - - - - - - - - - - 3
2. Young stems densely tawny-pubescent; leaves pubescent or scabrous above; secondary veins from the midrib reticulate, inconspicuous; styles unbranched; fruit subglobose, tawny-pubescent, up to 7(8) mm. long on pedicels 10–25 mm. long - - - 1. *africana*
– Young stems very sparsely white-pubescent or glabrous; leaves smooth and ± shiny above; secondary veins from the midrib parallel and running almost horizontally between it and the basal lateral nerves, conspicuous; styles shortly bifid; fruit ovoid, glabrous, usually 9–12 mm. long on pedicels 3–7 mm. long - - - - - - - - 2. *philippensis*
3. Styles 1–2-branched; stem pubescence (when present) tawny; fruit ovoid-ellipsoid, 7–10 mm. long; leaves usually ± crenate in the upper half; acumen short, up to $\frac{1}{7}(\frac{1}{5})$ of the blade in length; upper lateral nerves markedly more finely prominent than the midrib beneath and making a wide angle (greater than 45 degrees) with the midrib - - - - 3. *mildbraedii*
– Styles unbranched; stem pubescence (when present) whitish; fruit conical-ovoid, 5–6(7) mm. long; leaves entire, acumen long and usually $\frac{1}{4}-\frac{1}{3}$ of the blade in length; upper lateral nerves usually ± as thickly prominent as the midrib beneath and making an angle of less than 45 degrees with the midrib - - - - - - - 4. *gomphophylla*

1. **Celtis africana** Burm.f., Prodr. Fl. Cap.: 31 (1768). —Rendle in F.T.A. **6**, 2: 3 (1916). —Hauman in F.C.B. **1**: 43 (1948). —Keay in F.W.T.A. ed. 2, **1**: 592 (1958). —F. White, F.F.N.R.: 22 (1962). —Polhill in Kew Bull. **19**: 139 (1964). —von Breitenbach, Indig. Trees Southern Afr. **2**: 79, t. on page 81 (1965). —Polhill in F.T.E.A., Ulmaceae: 4 (1966). —Letouzey in Fl. Cameroun **8**: 14, t. 1 fig. 3 (1968). —Jacot Guillarmod, Fl. Lesotho: 161 (1971). —Palmer & Pitman, Trees S. Afr. **1**: 423, t. & photos (1972). —K. Coates Palgrave, Trees Southern Africa: 96, fig. 16 (1977). Type from S. Africa, being an illustration made from material collected in the Cape Province.
    Celtis rhamnifolia Presl., Bot. Burch.: 37 (1844) nom. nud. —N.E. Brown in F.C. **5**, 2: 518 (1925).
    Celtis kraussiana Bernh. in Flora **28**: 87 (1845). —Sim, For. Fl. Col. Cape Good Hope: 306, t. 134 (1907). —Engl., Pflanzenw. Afr.: 12 (1915). —Rendle in F.T.A. **6**, 2: 3 (1916). —Burtt Davy, Fl. Pl. Ferns Transv. **2**: 435, t. 66 (1932). —Peter in Fedde, Repert., Beih. **40**, 2: 64 (1932). —Hauman in F.C.B. **1**: 43 (1948). —Robyns, Fl. Sperm. Parc Nat. Alb. **1**: 43 (1948). —Brenan, Check-list For. Trees Shrubs Tang. Terr. **5**, 2: 624 (1949). —Andrews, Fl. Pl. Anglo-Egypt. Sudan **2**: 253 (1952). Type from S. Africa (Cape Province).
    Celtis burmannii Planch. in Ann. Sci. Nat., Bot., sér. 3, **10**: 296 (1848). Type from S. Africa (Cape Province).
    Celtis eriantha E. Mey. ex Planch. in Ann. Sci. Nat., Bot., sér. 3, **10**: 296 (1848). Types from S. Africa (Cape Province).
    Celtis opegrapha Planch. in Ann. Sci. Nat., Bot., sér. 3, **10**: 294 (1848). Type from S. Africa (Cape Province).
    Celtis vesiculosa Hochst. ex Planch. in Ann. Sci. Nat., Bot., sér. 3, **10**: 295 (1848). Type from Ethiopia.
    Celtis henriquesii Engl. in Not. Bot. Gart. Mus. Berl. **3**: 22 (1900). —Engl., Pflanzenw. Afr.: 12 (1915). Type from Angola.
    Celtis holtzii Engl. in Pflanzenw. Afr.: 12, t. 6 fig. E (1915). Type from Tanzania.
    Celtis kraussiana var. *stolzii* Peter in Fedde, Repert., Beih. **40**, 2: 64 (1932). —Brenan, Check-list For. Trees Shrubs Tang. Terr. **5**, 2: 624 (1949). Type from Tanzania.
    Celtis australis sensu A. Rich., Tent. Fl. Abyss. **2**: 257 (1851) non L.

A spreading tree to 30(35) m. tall, or a shrub, mostly deciduous, monoecious; bole slightly fluted, bark smooth, whitish-grey often pinkish-blotched; freshly cut wood unpleasant-smelling; young stems and branches densely tawny pubescent-tomentose. Leaves 3–9 × 2.5–5.5 cm. (juvenile leaves up to 13 × 7 cm.), broadly or narrowly ovate to ovate-lanceolate; apex acuminate; base rounded to cuneate and strongly asymmetrical; margin coarsely dentate-serrate in the upper two thirds, rarely crenate or subentire; lamina thinly or thickly chartaceous, ± scabrous, young foliage often densely tawny-pubescent on both sides, later glabrescent except on nerves below; 3-nerved from the base, basal lateral nerves extending well into upper half, upper lateral nerves 1–2 on each side of the midrib, all lateral nerves prominent below; petiole 2.5–5(10) mm. long; stipules linear to linear-obovate, 3–8 mm. long, pubescent. Inflorescences precocious; those cymes borne in the lower leaf axils and at the nodes below contain 3–many clustered male flowers on pedicels 1.5–5 mm. long; those cymes borne in the uppermost leaf axils contain 1–several bisexual flowers on pedicels 10–17 mm. long; intermediate cymules consisting of both male and bisexual flowers; axis and pedicels usually densely tawny-pubescent. Sepals 4–5, 1.5–2.5 mm. long, pubescent. Ovary densely pubescent; styles unbranched, 1.5–3.5(4) mm. long. Fruit orange, (4)5–7(8) × 3–6 mm. (dried), subglobose, less often ovoid-ellipsoid, pubescent; pedicels 10–25 mm. long.

Botswana. SE: Tlokweng, fl. 18.viii.1978, *Hansen* 3432 (BM; PRE; SRGH); Ootse Mt., fr. 1.i.1979, *Woollard* 495 (SRGH). **Zambia**. W: Solwezi Distr., fl. 26.vii.1964, *Fanshawe* 8848 (K). S: Chibila R., fr. x.1932, *Trapnell* 1127 (K). **Zimbabwe**. N: Mazowe Distr., Chipoli Farm, fl. 21.ix.1958, *Moubray* 19 (K; SRGH). W: Matobo Distr., Besna Kobila Farm, fr. xii.1953, *Miller* 1984 (K; PRE; SRGH). C: Rusape Distr., st. 7.xii.1954, *Munch* 427 (K; PRE; SRGH). E: Mutare Distr., Commonage, fl. 5.ix.1953, *Chase* 5062 (BM; K; PRE; SRGH). S: Mberengwa Distr., Buhwa Mt., st. 1.v.1973, *Pope, Biegel & Simon* 1039 (K; SRGH). **Malawi**. N: Misuku Hills, Wilindi Forest, fr. 12.xi.1958, *Robson & Fanshawe* 579 (K; SRGH). C: Dedza Distr., Ngoma, Chongoni Forest, fr. 16.xi.1970, *Salubeni* 1498 (K; SRGH). S: Thyolo (Cholo) Mt., fl. 22.ix.1946, *Brass* 17744 (BM; K; SRGH). **Mozambique**. MS: fr. 15.x.1954, *Chase* 5309 (BM; SRGH). GI: Lhanguene, st. 10.vii.1947, *Pedro & Pedrogão* 1430 (SRGH). M: Libombos (Lebombo) Mts., M'ponduine forest, fr. ii.1974, *Tinley* 3011 (K; SRGH).

Widespread in Africa from Nigeria to the Sudan and Arabia, and south to Angola and S. Africa (Cape). Not common but widely distributed at medium to higher altitudes in evergreen rainforests, wooded ravines, riverine fringes and on rocky outcrops in high rainfall areas, sometimes on termite mounds (occurs in coastal forests in S. Africa); 0–2000 m.

2. **Celtis philippensis** Blanco, Fl. Filip.: 197 (1837). —Leroy in Fl. Madag., fam. 54: 3, t. 3 fig. 1–4 (1952). —Berhaut, Fl. Seneg.: 144 (1954). —Letouzey in Fl. Cameroun 8: 26–32, tabs. 2, figs. 1–2; t. 4, fig. 1. Type from Philippines.

Celtis wightii Planch. in Ann. Sci. Nat., Bot., sér. 3, **10**: 307 (1848). —Polhill in Kew Bull. **19**: 141 (1964); in F.T.E.A., Ulmaceae: 9 (1966). Types from India & Sri Lanka.

Celtis mauritiana Planch. in Ann. Sci. Nat., Bot., sér. 3, **10**: 307 (1848). —Engl., Pflanzenw. Ost-Afr. **C**: 160 (1895). —Rendle in F.T.A. **6**, 2: 9 (1916). —Peter in Fedde, Repert., Beih. **40**, 2: 64 (1932). —Brenan, Check-list For. Trees Shrubs Tang. Terr. **5**, 2: 624 (1949). Type from Mauritius.

Celtis prantlii Engl. in Not. Bot. Gart. Mus. Berl. **3**: 23 (1900); in Mildbr. Wiss. Ergebn. Deutsch. Zentral-Afr.-Exped. 1907–1908, **2**: 179 (1911). —Engl., Pflanzenw. Afr.: 12 (1915). —Rendle in F.T.A. **6**, 2: 8 (1916). —Hutch., Dalz. & Moss in F.W.T.A. **1**: 423 (1928). —Hauman in F.C.B. **1**: 43 (1948). —Robyns, Fl. Sperm. Parc Nat. Alb. **1**: 44 (1948). Type from W. Africa.

Celtis insularis Rendle in Journ. Bot. **53**: 297 (1915), nom. illegit. Type as for *C. prantlii*.

Celtis brownii Rendle in Journ. Bot. **53**: 298 (1915); in F.T.A. **6**, 2: 10 (1916). —Keay in F.W.T.A. ed. 2, **1**: 592 (1958). Type from Uganda.

Celtis scotellioides A. Chev. in Bull. Soc. Bot. Fr. **61**, Mém. 8e: 299 (1917). Type from Ivory Coast.

Celtis rendleana G. Tayl. in Exell, Cat. Vasc. Pl. S. Tomé: 302 (1944). Type from Angola.

Evergreen much-branched tree 5–20 m. tall, monoecious; bole often with short buttresses, bark smooth grey, wood white; young stems and branches very sparsely white-pubescent, glabrescent. Leaves (5)6.5–12(17.5) × 2.5–4.5(8) cm., elliptic to elliptic-ovate or elliptic-oblong; apex with a wide (usually short) acumen and mucronate tip; base slightly asymmetrical and broadly cuneate to rounded or subcordate; margin entire or (rarely in Flora Zambesiaca region) coarsely dentate in the upper half; lamina thinly coriaceous, pale green, punctate but smooth and ± shiny above, glabrous; 3-nerved from the base, the basal lateral nerves extending almost to the apex, as strongly prominent beneath as the midrib, upper lateral nerves 1–2(3) on each side of the midrib, rather fine and inconspicuous; secondary venation fairly closely parallel, ± horizontal between the midrib and basal lateral nerves. Petiole 4–10(16) mm. long; stipules 3–7 mm. long, lanceolate, shortly produced below the point of attachment, ± pubescent. Inflorescences in leaf axils towards the ends of branches; lower inflorescences 5–30 mm. long, containing many crowded male flowers, these sessile or with pedicels to 2 mm. long, and a few female and bisexual flowers at apices of branches of these inflorescences, their pedicels usually longer; upper inflorescences usually short with several bisexual flowers. Sepals 1.5–2.5 mm. long, pubescent. Ovary ± glabrous, with a basal ring of long hairs; styles very shortly bifid, (1)1.5–2(3.5) mm. long. Fruits red, 9–12 × 7–10 mm., ovoid, glabrous; pedicels 3–7 mm. long.

Mozambique. MS: Condué R., 18°05'S, 35°02'E, fr. 11.x.1961, *Gomes e Sousa* 4709 (COI; K); Siluva (Silura) Hills, fr. 1924, *Honey* 663a (K).
Recorded from West Africa to Ethiopia, and south to east tropical Africa and Angola; also in Madagascar and Mascarene Is., and from tropical Asia to Australia. In low altitude evergreen forest, swamps and riverine forest; c. 300 m.

3. **Celtis mildbraedii** Engl., Bot. Jahrb. **43**: 309 (1909); in Mildbr. Wiss. Ergebn. Deutsch. Zentral-Afr.-Exped. 1907–1908, **2**: 180, t. 16, fig. E (1911); Pflanzenw. Afr.: 14 (1915). —Hauman in F.C.B. **1**: 45 (1948). —Keay in F.W.T.A. ed. 2, **1**: 592 (1958). —Polhill in Kew Bull. **19**: 140 (1964); in F.T.E.A., Ulmaceae: 7 (1966). —Letouzey in Fl. Cameroun **8**: 33, tabs. 3, 4 fig. 2 & t. 5 fig. 4 (1968). —Palmer & Pitman, Trees of S. Afr. **1**: 427, t. & photo (1972). —K. Coates Palgrave, Trees Southern Africa: 97 (1977). Type from Zaire.
Celtis soyauxii sensu the following authors: Engl. in Not. Bot. Gart. Mus. Berl. **3**: 23 (1900) pro parte. —Rendle in F.T.A. **6**, 2: 5 (1916). —Battiscombe, Trees & Shrubs Kenya: 84 (1936). —Brenan, Check-list For. Trees Shrubs Tang. Terr. **5**, 2: 624 (1949). —Andrews, Fl. Pl. Anglo-Egypt. Sudan **2**: 251 (1952). —Keay in F.W.T.A. ed. 2, **1**: 592 (1958); excluding the type which is *C. zenkeri* Engl.
Celtis usambarensis Engl., Bot. Jahrb. **43**: 309 (1909); Pflanzenw. Afr.: 14 (1915). —Peter in Fedde, Repert., Beih. **40**, 2: 65 (1932). Type from Tanzania.
Celtis compressa A. Chev., in Bull. Soc. Bot. Fr. **61**, Mém. 8e: 298 (1917). Type from Ivory Coast.
Celtis franksiae N.E. Brown in F.C. **5**, 2: 517 (1920). —Henkel, Woody Pl. Natal: 106 (1934). —von Breitenbach, Indig. Trees S. Afr. **2**: 82 (1965). Type from S. Africa (Natal).
Celtis bequaertii De Wild. in Rev. Zool. Bot. Afr. 9, Suppl. Bot.: 2 (1921); Pl. Bequaert.: 189 (1922). Type from Zaire.
Celtis dubia De Wild. in Rev. Zool. Bot. Afr. 9, Suppl. Bot.: 5 (1921); Pl. Bequaert.: 190 (1922). Type from Zaire.

Evergreen or deciduous tree 3–40 m. tall, monoecious; branches often drooping; bole with buttresses, bark pale smooth or scaling in small disks; young twigs tawny-pubescent. Leaves (7.5)9–15 × 4–5(8) cm., elliptic to elliptic-obovate; apex long acuminate, ± mucronate; base cuneate, slightly asymmetrical; margin obscurely crenate to coarsely dentate in the upper half; lamina chartaceous to thinly coriaceous, venation as in *C. gomphophylla* but upper lateral nerves beneath less strongly prominent than the midrib, making an angle of more than 45 degrees with the midrib; very young leaves sparsely tawny-pubescent, soon glabrous; petiole 3–10 mm. long; stipules 4–5 mm. long, lanceolate, tawny-pubescent. Cymes 4–15 mm. long, of many, rarely few, often crowded male flowers with pedicels up to 2 mm. long, female and bisexual flowers 1 or few at the apex of the cyme; uppermost cymules of the inflorescence with several bisexual flowers. Sepals 5, 1.5–2 mm. long, pubescent. Ovary often with a ring of sparse hairs at the base otherwise subglabrous; styles once or twice-branched, c. 5 mm. long. Fruits red, 7–10 × 5–6 mm., ovoid-ellipsoid, 4-ribbed when dry, glabrous; pedicels 8–13(20) mm. long.

Zimbabwe. E: Chipinge Distr., Chirinda forest, fl. & fr. x.1962, *Goldsmith* 197/62 (BM; COI; K; SRGH); same locality, fr. x.1962, *Goldsmith* 231/62 (K; SRGH). **Mozambique**. Z: Morrumbala, st. 30.v.1974, *Bond* W.394 (SRGH). MS: Mt. Zembe, E. slopes, st. 19.vii.1970, *Müller & Gordon* 1336 (K; SRGH); Inhaminga, *Xylia* thicket, st. ix.1972, *Earle* 5-P69A (SRGH).
Recorded from West Africa to the Sudan and south through Tanzania, Zaire and Angola to South Africa (Natal). Uncommon, in low to medium altitude, evergreen rainforest; 200–1100 m.

4. **Celtis gomphophylla** Bak. in Journ. Linn. Soc., Bot. **22**: 521 (1887). —Leroy in Fl. Madag., fam. 54: 6, t. 3, fig. 5–8 (1952). —Letouzey in Fl. Cameroun **8**: 39, tabs. 2 fig. 4 & t. 5 figs. 1–2 (1968). —K. Coates Palgrave, Trees Southern Africa: 97 (1977). TAB. 3. Type from Madagascar.
Celtis durandii Engl. in Not. Bot. Gart. Mus. Berl. **3**: 22 (1900); in Mildbr. Wiss. Ergebn. Deutsch. Zentral-Afr.-Exped. 1907–1908, **2**: 179 (1911); Pflanzenw. Afr.: 12, t. 6 fig. D (1915). —Rendle in F.T.A.: **6**, 2: 4 (1916). —Peter in Fedde, Repert., Beih. **40**, 2: 65 (1932). —Hauman in F.C.B. **1**: 42 (1948). —Robyns, Fl. Sperm. Parc Nat. Alb. **1**: 43 (1948). —Brenan, Check-list For. Trees Shrubs Tang. Terr. **5**, 2: 624 (1949). —Keay in F.W.T.A. ed. 2, **1**: 592 (1958). —F. White, F.F.N.R.: 431 (1962). —Polhill in Kew Bull. **19**: 140 (1964); in F.T.E.A., Ulmaceae: 5 (1966). —Palmer & Pitman, Trees of S. Afr. **1**: 427, t. & photo (1972). Types from Zaire and Tanzania.

Tab. 3. CELTIS GOMPHOPHYLLA. 1, flowering branchlet (×⅔); 2, male flower with one sepal and stamen removed (× 8); 3, same with aborted ovary more developed (× 8); 4, sepal and stamen (× 12), 1–4 *Kakoire* 79; 5, female flower (× 8); 6, longitudinal section of female flower (× 6), 5–6 *Koritschoner* 1574; 7, fruiting branchlet (×⅔), *Wallace* 1201. Drawn by M.E. Church. From F.T.E.A.

Celtis ugandensis Rendle in Journ. Bot. **44**: 341 (1906). Type from Uganda.
  Celtis dioica S. Moore in Journ. Linn. Soc., Bot. **40**: 204 (1911). Type: Zimbabwe, Gazaland,
Chirinda Forest, 3–4000 ft., fl. 8.x.1906, *Swynnerton* 108 (BM, holotype; K).
  Celtis durandii var. *ugandensis* (Rendle) Rendle in F.T.A. **6**, 2: 5 (1916). —Hauman in F.C.B. **1**:
43 (1948). —Robyns, Fl. Sperm. Parc Nat. Alb. **1**: 43 (1948).

Deciduous tree to 3–30(60) m. tall, monoecious or dioecious; bole often fluted or
buttressed, bark smooth light grey, wood unpleasant smelling; young stems and branches
whitish-pubescent. Leaves (5)6–16 × 2–5(7) cm., ovate-elliptic to oblong-elliptic; apex
long-acuminate; base cuneate to rounded, asymmetrical; margin entire or with a few
coarse teeth, (juvenile foliage up to 21 × 9 cm., with apex hardly acuminate and margin
coarsely dentate in upper half); lamina membranous-chartaceous, glabrescent, often
scabrid, 3-nerved from the base with the basal lateral nerves not or hardly extending into
the upper half; upper lateral nerves (2)3–6 on each side of the midrib, prominent above,
more strongly so below, usually making an angle of less than 45 degrees with the midrib;
petiole 4–8 mm. long; stipules 2–6 mm. long, linear to linear-oblong, whitish pubescent,
caducous. Flowers precocious; male flowers in numerous, crowded, few–many-flowered
cymes, pedicels 3–7 mm. long; female and bisexual flowers few or solitary, axillary or at
nodes below, pedicels often longer. Sepals 4–5, 1.2–2 mm. long, pubescent. Ovary
± pubescent or glabrous; styles unbranched, 2–2.5 mm. long. Fruit dark yellowish, 4–6(7)
× 3–5 mm., conical-ovoid, often 4-angled when dry, glabrous, pedicel 3–10 mm. long.

Zambia. N: Sunzu Hill, fr. 31.iii.1960, *Fanshawe* 5600 (K). **Zimbabwe**. E: Chimanimani Distr.,
Gungunyana For. Res., fr. xi.1961, *Goldsmith* 96/61 (K; SRGH); Chipinge Distr., Chirinda Forest, fl.
x.1965, *Goldsmith* 25/65 (K; SRGH). **Malawi**. N: Misuku Distr., Mughama, st. 28.iv.1963, *Chapman*
1958 (SRGH). C: Nkhota-Kota Distr., Ntchisi Forest, st. 10.vii.1960, *Chapman* 811 (SRGH). S:
Mangochi Distr., Mangochi Mt., st. 21.xii.1963, *Chapman* 2149 (SRGH). **Mozambique**. MS: Beira
Distr., Cheringoma Coast, Waroa Forest, st. v.1973, *Tinley* 2922 (K; SRGH).
  Also recorded from Nigeria, São Tomé, Zaire, Angola and S. Africa (Natal and Cape). Uncommon
in low to medium altitude, mixed evergreen rainforest and riverine forest (occurs in coastal forest in
S. Africa); 700–1400 m.

155. CANNABACEAE

By C.M. Wilmot-Dear

Herbs, annual or perennial, erect or climbing, without latex. Leaves alternate or
opposite, petiolate, simple and undivided to palmately lobed or digitately compound;
stipules free or fused. Inflorescences numerous and axillary, flowers dioecious and
wind-pollinated. Male inflorescences paniculate, flowers with perianth uniseriate, 5-
lobed, imbricate; stamens 5, opposite the perianth lobes, anthers straight, erect in bud,
2-thecous, dehiscing at first by apical pores but soon also lengthwise; pistillode absent.
Female inflorescences strobilate, flowers ± sessile, crowded, tightly enclosed or loosely
subtended by small or large persistent bracteoles; bracts present; perianth membranaceous,
entire, investing the ovary; ovary superior, sessile, 1-locular; ovule 1, pendulous,
anatropous; style terminal, short; stigmas 2, long filiform. Fruit an achene covered by the
persistent perianth; endosperm sparse, fleshy and oily; embryo curved or spirally coiled.

A family of 2 genera native to temperate parts of the northern hemisphere, 1
naturalised in Africa. The number of species in both genera is disputed: *Cannabis* is being
variously considered as monotypic or with up to 3 species; *Humulus* is usually held to have
2 species but some authors recognise 3–4 species.
  Both genera are of economic importance: *Cannabis* as a source of hemp fibre and an
intoxicant resin (the hallucinogenic drug known as "bhang", "dagga", "hashish",
"marijuana", "indian hemp" and "pot"); *Humulus* as a flavouring in beer. (See N.G. Miller
in J. Arn. Arb. **51**: 185–203 (1970) for a detailed account and extensive bibliography).

CANNABIS L.

Cannabis L., Sp. Pl. **2**: 1027 (1753); Gen. Pl. ed. 5: 453 (1754).

Erect annual aromatic herbs, dioecious with male and female plants dimorphic, or
rarely monoecious; indumentum of most parts consisting of minute appressed swollen-

based hairs; leaves alternate or opposite at the stem base, palmately lobed or digitately compound, leaflets uneven in size, with serrate margins; stipules lateral, persistent. Male inflorescences laxly cymose-paniculate, bristly-hairy, exceeding the leaves but bearing a few scattered leaves; flowers numerous, shortly pedicellate; perianth lobes free, boat-shaped, spreading or reflexed; stamens pendulous at maturity, filaments short; pistillode 0. Female inflorescences short, compact, not exceeding the leaves; flowers fewer, paired, each subtended by a stipule-like bract and a small green "bracteole" or "calyx" which completely envelopes the ovary and loosely encloses the mature fruit, this enveloping bracteole forming a basally swollen tubular sheath narrowly attenuate at the apex and covered with fine hairs and shortly-stalked or sessile, resinous glands; perianth membranous, undivided, tightly enveloping the ovary and mature fruit (often reduced or absent in cultivated forms), marbled with light and dark areas; stigma-branches densely pubescent, caducous. Fruit a globular to ovoid achene tightly covered by the thin crustaceous perianth, the reticulate venation of the surface visible beneath the perianth. Embryo strongly curved, cotyledons fleshy.

Easily identified by the light and dark patterning of the perianth layer (where this is present) surrounding the fruit.

Cannabis has been treated variously as comprising 3 species (Anderson in Bot. Mus. Leafl., Harv. Univ. **28**: 61–69 (1980)) or as 1 very variable species in which 4 infraspecific taxa can be recognised (Small & Cronquist in Taxon **25**: 405–435 (1976)).

The inherent variation within this genus has, by artificial selection (for production of fibre, oil or intoxicating resin) followed by naturalisation, crossbreeding and recombination of characters, given rise to a reticulate pattern of variation where, primarily in the female plants, several extreme forms exist, but where a continuous range of intermediates is also present. Variation in male plants is far less extreme. However, while it may be possible to recognise different groups within the female plants it remains difficult or impossible to assign male plants to such groupings, and for this reason the genus is best considered as comprising a single variable species.

Cannabis sativa L., Sp. Pl. **2**: 1027 (1753). —Engl., Pflanzenw. Ost-Afr. **C**: 162 (1895). —Engl., Mon. Afrik. Pflanzen. 1 (Moraceae): 44 (1898). —Hiern, Cat. Afr. Pl. Welw. **1**, 4: 994 (1900). —Rendle in F.T.A. **6**, 2: 16 (1916). —Burtt Davy, Fl. Pl. Ferns Transv. **2**: 445 (1932). —Hauman in F.C.B. **1**: 176 (1948). —Henderson & Anderson in Mem. Bot. Surv. S. Afr. **37**: 70 (1966). —Purseglove, Tropical Crops, Dicots. **1**: 40, fig. 4 (1968). —Verdc. & Trump, Common Poisonous Pl. E. Afr.: 96 (1969). —Miller in Journ. Arn. Arb. **51**: 185–203 (1970). —Stearn in Joyce & Curry, The Botany and Chemistry of Cannabis; 1–10, fig. 1–7 (1970). —Jacot Guillarmod, Fl. Lesotho: 162 (1971). —Stearn in Bot. Mus. Leafl., Harv. Univ. **23**: 325–336 (1974). —Verdcourt in F.T.E.A., Cannabaceae: 1 (1975). —Small & Cronquist in Taxon **25**, 4: 405–435 (1976). —Emboden in Taxon **26**, 1: 110 (1977). —Anderson in Bot. Mus. Leafl., Harv. Univ. **28**: 61–69 (1980). TAB. **4**. Type a female specimen in Hort. Cliff.

Cannabis indica Lam., Dict. Encycl. **1**: 697 (1785). Type from India.

Cannabis sativa var. *indica* (Lam.) Wehmer, Die Pflanzenstoffe: 157 (1911) but see note in Small & Cronquist (1976).

Cannabis sativa var. *spontanea* Vavilov in Trudy Prikl. Bot. Selekc. **13** (suppl. 23): 148 (1922). Type from USSR.

Cannabis ruderalis Janischevsky in Uchen. Zapiski Univ. Saratov **2**, 2: 14 (1924). Type from USSR.

An erect, rank smelling, annual herb up to 2(4.5) m. tall; male plants taller and more slender with longer narrower leaflets than the female, dying soon after flowering; female plants short, more robust, with densely leafy inflorescences, the plants living for several months after pollination. Indumentum of most parts consisting of minute appressed swollen-based hairs. Stems simple or branched, leafy, angular, often with hollow internodes. Leaves 3–7(11)-digitately-foliolate; petiole 2–6 cm. long; stipule to 1.4 cm. long, linear, acute; leaflets sessile, 2.5–15 × 0.35–2 cm., narrowly lanceolate, tapering acuminate at the apex, narrowly cuneate at the base; margin coarsely dentate to serrate-biserrate; lamina membranous-chartaceous, shortly coarsely-hairy and yellow-glandular on both surfaces, penninerved, midrib prominent beneath. Male and female flowers rarely both on 1 plant, if so then 1 predominating. Male inflorescences numerous, loosely cymose-paniculate, sparsely leafy, up to 20(30) cm. long, few to more than 20-flowered; bracts to c. 15 mm. long, bristly-hairy. Male flowers small, pedicellate, regular; pedicels to 7 mm. long; perianth lobes free, greenish or white, 3–4 × 1 mm., oblong-elliptic, boat-shaped, spreading or reflexed, appressed-pubescent outside; stamens at length pendulous, filaments 0.3–1 mm. long, anthers 3–4 mm. long. Female inflorescence short, crowded or strobilate, few-flowered, densely leafy; bracts often shorter than in male flowers. Female flowers ± sessile; enveloping bracteole 2–8 mm. long, green; perianth

Tab. 4. CANNABIS SATIVA. 1, male flowering shoot (×⅔), *Ward* 6086; 2, male inflorescence (× 3); 3, male flower (× 6); 4, stamen (× 6), 2–4 *Semsei* 1667; 5, female inflorescence (× 4); 6, female flower, with bracteole (× 6); 7, female flower (× 6), 5–7 *Holst* 2685; 8, fruit, enveloped by bracteole (× 4); 9, achene (× 4), 8–9 *Eggeling* 1266. From F.T.E.A.

thin, undivided, enveloping ovary and mature fruit (often reduced or absent in cultivated forms); ovary sessile, c. 1 mm. in diam., ± globose; stigma branches (1)2–5 mm. long, pubescent, caducous. Fruit 3–4 × 2–3.5 mm., globular to ovoid, surface uniformly coloured, pale with a prominent reticulate pattern of venation or, where a persistent perianth is present, shiny brownish or greyish, mottled with a light and dark marbled pattern, venation visible beneath.

Botswana. SE: Mahalapye, female fl. 20.xii.1911, *Rogers* 6083 (SRGH). **Zambia**. B: Senanga, fl. viii.1933, *Trapnell* 1265 (K). W: Kitwe, grown in forest nursery, female fl. & fr. 22.iii.1958, *Fanshawe* 4364 (K). C: Chakwenga Headwaters, 100–120 km. E. of Lusaka, abandoned village, male fl. 27.iii.1965, *Robinson* 6560 (K; SRGH). S: Mazabuka Distr., Shamonyemba Village, between Magoye and Kaleya rivers, female fl. 16.vii.1963, *van Rensberg* 2342 (K; SRGH). **Zimbabwe**. W: Bulilima-Mangwe Distr., Embakwe Mission, female fl. xii.1941, *Feiertag* in GHS 45526 (PRE; SRGH). C: Marondera Distr., Marandellas Grassland Res. Station, st. 4.i.1964, *Corby* 1057 (K; LISC; SRGH). **Malawi**. C: Dedza Distr., Chongoni For. Res., female fl. & fr. 5.iii.1968, *Salubeni* 966 (K; SRGH). S: Mulanje Distr., near Litchenya Forestry Hut, fr. 29.iii.1960, *Phipps* 2783 (K; SRGH). **Mozambique**. N: Nampula, Meconta, male fl. iv.1937, *Torre* 1380 (COI; LISC). MS: Vila Machado, serra de Chiluro, female fl. 16.iv.1948, *Mendonça* 3982 (LISC). GI: Inharrime (Nhacoongo), st. 25.x.1947, *Barbosa* 537 (SRGH).

A native of Central Asia, widely cultivated and naturalised throughout the world; naturalised in Africa as a weed of old cultivations; also widely cultivated for the intoxicant resin, known locally as "dagga".

Most collections from South Africa can be assigned to one or other of 3 varieties:
1. var. *sativa* — plants tall, sparsely laxly branched with large leaves consisting of 5–7, ± narrowly lanceolate leaflets and large fruits with a poorly developed or readily deciduous perianth; modified by selective cultivation for fibre and seed-oil.
2. var. *indica* (Lam.) Wehmer — plants more robust, somewhat shorter with many crowded branches; leaflets usually more numerous, oblanceolate and relatively weaker; fruits large with a poorly developed or readily deciduous perianth; modified by cultivation for intoxicant resin.
3. var. *spontanea* Vavilov — plants small, not or little branched; leaves small with few elliptic leaflets; fruit smaller with a persistent well developed perianth; plants either little modified by cultivation or having reverted due to naturalisation and interbreeding of cultivated forms.
In the Flora Zambesiaca area however, most specimens seen possess combinations of characters so intermediate between these 3 varieties that it is best to treat this species as *C. sativa* L. sens. lat.

156. MORACEAE

By C.C. Berg*

Trees, shrubs or herbs, dioecious or monoecious; sap milky, sometimes watery (but not turning black). Leaves spirally or distichously arranged, sometimes subopposite or subverticillate, entire or sometimes pinnately or palmately incised, stipulate. Inflorescence bisexual or unisexual, spicate, globose, clavate- or discoid-capitate, urceolate, sometimes uniflorous. Staminate flowers with 2–6 tepals or perianth lacking; stamens 1–4. Pistillate flowers with 2–6 tepals or perianth lacking; pistil 1; ovary free or adnate to the perianth; stigmas 1 or 2, ovule 1, apically attached. Fruit achene-like, drupaceous (dehiscent or not), or forming a drupaceous whole with the fleshy perianth or with the fleshy receptacle as well. Seed large and without endosperm or small with endosperm, embryo various.

A family of c. 50 genera and some 1100–1150 species, the majority tropical with c. 625 species in Asia and Australasia, c. 300 species in the Neotropics, and c. 200 species in the African region.

Species introduced into the Flora Zambesiaca area include:
Artocarpus altilis (Parkinson) Fosberg (*A. communis* J.R. & G. Forster). "Bread Fruit". Native of tropical Asia. Tree; leaf mostly 30–50 × 15–20 cm., pinnately incised; inflorescence borne in the leaf axils.
Artocarpus heterophyllus Lam. "Jack Fruit". Native of tropical Asia. Tree; leaf 10–20 × 6–10 cm., entire; inflorescences borne on the trunk and the main branches.

*Dorstenia by M.E.E. Hijman. Both authors received grants from the Netherlands Organization for Advancement of Pure Research (ZWO) to carry out this study.

14 156. MORACEAE

Cultivated species of *Ficus, Morus* and *Maclura* are mentioned below under these genera.

1. Plants herbaceous, often somewhat succulent - - - - - **9. Dorstenia**
   - Plants woody - - - - - - - - - - - - - - 2
2. Flowers concealed within a hollow, fleshy nearly closed receptacle (fig) with a small apical mouth (ostiole); glandular, ± waxy spots on the leaf beneath at the base of the midrib or in the axils of at least the main basal lateral veins (TAB. 16, figs. 22–24) - - **10. Ficus**
   - Flowers borne on the inner surface of a ± flattened or concave, open receptacle, or in spikes or globose heads; glandular spots absent on the leaf - - - - - - 3
3. Stipules fully amplexicaul (leaving annular scars) - - - - - - 4
   - Stipules semi-amplexicaul to lateral - - - - - - - - - 6
4. Stipules connate; inflorescences bisexual; staminate flowers without perianth **8. Trilepisium**
   - Stipules free; inflorescences unisexual, or if bisexual, then the staminate flowers with a perianth - - - - - - - - - - - - - - 5
5. Inflorescences globose- to obovoid-capitate - - - - - **5. Treculia**
   - Inflorescences discoid- to turbinate-capitate or uniflorous - - - **7. Bosqueiopsis**
6. Plants armed with spines - - - - - - - - - - **3. Maclura**
   - Plants without spines - - - - - - - - - - - 7
7. Inflorescences bisexual - - - - - - - - - - **9. Dorstenia**
   - Inflorescences unisexual - - - - - - - - - - 8
8. Staminate inflorescences discoid-capitate and involucrate; pistillate inflorescences uniflorous, involucrate, flower adnate to the receptacle - - - - - **6. Antiaris**
   - Staminate inflorescences spicate; pistillate inflorescences globose-capitate or uniflorous with the flower free - - - - - - - - - - - - - 9
9. Leaf beneath densely puberulous on the vein-reticulum; both staminate and pistillate inflorescences spicate - - - - - - - - - - - **2. Milicia**
   - Leaf beneath sparsely puberulous to glabrous; staminate inflorescence spicate; pistillate inflorescences subglobose or uniflorous - - - - - - - - 10
10. Leaf subtriplinerved; peduncle of the staminate inflorescence 0.3–2 cm. long; pistillate inflorescences several-flowered - - - - - - - - **1. Morus**
    - Leaf pinnately veined; peduncle of staminate inflorescences up to 0.2 cm. long; pistillate inflorescences uniflorous - - - - - - - - **4. Streblus**

1. MORUS L.

Morus L., Sp. Pl.: 986 (1753); Gen. Pl. ed. 5: 424 (1754). —C.C. Berg in Bull. Jard. Bot. Brux. **47**: 335 (1977); in F.T.E.A., Moraceae: 2 (1989).

Subgen. **Afromorus** A. Chev. ex Lcroy, Rev. Int. Bot. Appl. Agric. Trop. **29**: 482 (1949); in Bull. Mus. Nation. Hist. Nat. Paris, sér. 2, **21**: 732 (1949). —C.C. Berg in Bull. Jard. Bot. Brux. **47**: 335 (1977).

Trees, dioecious, the shoot apices shed. Leaves distichously arranged, subtriplinerved; stipules lateral, free. Inflorescences on the lower leafless nodes of lateral branches (or in the leaf axils), bracteate. Staminate inflorescences spicate; tepals 4, basally connate; stamens 4, inflexed in bud; pistillode present. Pistillate inflorescences capitate; tepals 4, basally connate; ovary free; stigmas 2, filiform, subequal in length. Fruiting perianth enlarged, fleshy, greenish to yellow; fruit free, somewhat drupaceous. Seed small, with endosperm; cotyledons thin, equal, plane.

A genus of 10–15 species in temperate to subtropical regions of the Old and New Worlds, with only one species native in Africa. Two species are cultivated in southern Africa: 1. *Morus alba* L. (the White Mulberry) with the leaf lamina smooth above, style absent, stigmas 1–1.5 mm. long and sessile and the fruiting perianth mostly white. 2. *Morus australis* Poir. (*M. indica* sensu auctt., non L.), the cultivated mulberry, widely grown in the Flora Zambesiaca area; leaf lamina usually scabrous above; style present, fruiting perianth dark red to black.

Morus mesozygia Stapf in Journ. de Bot. (Paris), sér. 2, **2**: 99 (1909). —Rendle in F.T.A. **6**, 2: 21 (1916). —Hauman in F.C.B. **1**: 55 (1948). —Keay in F.W.T.A. ed. 2, **1**: 594 (1958). —Gomes e Sousa, Dendrol. Moçamb. **1**: 194, t. 13 (1967). —K. Coates Palgrave, Trees Southern Africa: 100 (1977). —C.C. Berg in Bull. Jard. Bot. Brux. **47**: 337, fig. 16 (1977); in Fl. Cameroun **28**: 6, t. 1 (1985); in F.T.E.A., Moraceae: 2 (1989). TAB. 5. Lectotype from Ivory Coast, chosen by Berg et al. in Fl. Cameroun **28**: 8 (1985).
   Celtis lactea Sim, For. Fl. Port. E. Afr.: 97, t. 91 (1909). —Rendle in F.T.A. **6**, 2: 4 (1916). Type: Mozambique, Quissico, *Sim* 5299 (K, holotype).
   Morus lactea (Sim) Mildbr. in Notizbl. Bot. Gart. Berl. **8**: 243 (1922). Type as above.
   Morus mesozygia var. *lactea* (Sim) A. Chev. in Rev. Bot. Appliq. **29**: 72 (1949). Type as above.

Tab. 5. MORUS MESOZYGIA. 1, leafy twig with staminate inflorescences, *Simao* 14; 2, leafy twig with pistillate inflorescences, *Fanshawe* 9319; 3, leafy twig with infructescences, *Simao* 233; 4, staminate flower and bracts; 5, pistillode, 4–5 *Andrada* 1447; 6, pistillate flower and bracts, *Gomes e Sousa* 1862; 7, pistillate flower in fruit, *Espirito Santo* 1961; 8, 9, fruits; 10, seed; 11, embryos, 8–11 *Simao* 233. Drawn by E.H. Hupkens van der Elst and W. Scheepmaker.

Tree up to 40 m. tall. Leaf lamina elliptic to oblong, ovate or subobovate, 3–13 × 2–8 cm., chartaceous to subcoriaceous, apex acuminate to subacute, base cordate to obtuse, margin crenate to serrate; superior surface pubescent on the main veins, inferior surface pubescent in the axils of the lateral veins; lateral veins 4–7 pairs, the basal pair strong, the others much less distinctive, arising from the upper part of the midrib, tertiary venation partly scalariform; petiole 0.5–2 cm. long; stipules c. 0.5 cm. long, caducous. Staminate inflorescences: spike 1–2.5 cm. long, c. 0.8 cm. in diam.; peduncle 0.3–2 cm. long. Pistillate inflorescences with 5–15 flowers c. 0.5 cm. in diam., (up to c. 1 cm. in diam. in fruit); peduncle 0.4–2 cm. long; stigmas 3–5 mm. long. Fruit ellipsoid to subglobose, ± compressed, 5 × 3–5 mm.

Zambia. N: Mbala, Lunzua Power Station, 13.iv.1962, *Lawton* 844 (FHO; NDO). W: Mpongwe, 12.ix.1965, *Fanshawe* 9319 (FHO; K; SRGH). **Malawi**. S: Mtemangokwe R., Mangochi (Fort Johnston), 11.x.1954, *Jackson* 1372 (FHO; K). **Mozambique**. N: Niassa, Mandimba, *Hornby* 2428 (PRE). Z: Namacurra, c. 57 km. de Nicuadala, estrada para Campo, 2.xi.1966, *Torre & Correia* 14381 (LISC; WAG). MS: Manica, Moribane, 4 km. on the road to Sanguene, 5.x.1953, *Pedro* 4220 (K; LMA; PRE). GI: Gaza, Chipenhe, estrada para Xai-Xai proximo de Mainguelane, 13.x.1957, *Barbosa & Lemos* 8027 (K; LISC; LMA; UC). M: Siluvu Hills, Maputo, 4.xi.1927, *Earthy* s.n. (PRE).

Also from Senegal to SW. Ethiopia, southwards to NW. Angola and S. Africa (N. Natal). Evergreen or semi-deciduous forest and riverine vegetation; 0–1600 m.

2. MILICIA Sim

Milicia Sim, For. Fl. Port. E. Afr.: 97 (1909). —C.C. Berg in Bull. Jard. Bot. Brux. **52**: 226 (1982); in F.T.E.A., Moraceae: 4 (1989).
Maclura Nuttall sect. *Chlorophora* (Gaud.) Baill., Hist. Pl. **6**: 193 (1875–76) pro parte. —Corner in Gard. Bull., Singapore **19**: 236 (1962).

Dioecious trees. Leaves distichous, pinnately veined, stipules semi-amplexicaul, free. Inflorescences in the leaf axils, spicate, bracteate. Staminate flowers: tepals 4, basally connate; stamens 4, inflexed in bud; pistillode present. Pistillate flowers numerous; tepals 4, basally connate; ovary free, stigmas 2, filiform, markedly unequal in length. Fruiting perianth enlarged, ± fleshy, greenish; fruit free, somewhat drupaceous. Seed small, with endosperm, cotyledons thin, equal, plane.

A tropical African genus of 2 species.

Milicia excelsa (Welw.) C.C. Berg in Bull. Jard. Bot. Brux. **52**: 227 (1982); in Fl. Cameroun **28**: 9, t. 2 (1985); in F.T.E.A., Moraceae: 4 (1989). TAB. **6**. Type from Angola.
Morus excelsa Welw. in Trans. Linn. Soc., Bot. **27**: 69, t.23 (1869). Type as above.
Maclura excelsa (Welw.) Bur. in DC., Prodr. **17**: 231 (1873). —Corner in Gard. Bull., Singapore **19**: 257 (1962). Type as above.
Chlorophora excelsa (Welw.) Benth. & Hook. f., Gen. Pl. **3**, 1: 363 (1880). —Rendle in F.T.A. **6**, 2: 22 (1916). —Hauman in F.C.B. **1**: 56 (1948). —Keay in F.W.T.A. ed. 2, **1**: 595 (1958). —Gomes e Sousa, Dendrol. Moçamb., Estudo Geral, **1**: 196, t. 14 (1966). —K. Coates Palgrave, Trees Southern Africa: 101 (1977). —C.C. Berg in Bull. Jard. Bot. Brux. **47**: 349, t. 19 (1977). Type as above.
Chlorophora tenuifolia Engl., Bot. Jahrb. **20**: 139 (1894). Type from São Tomé.
Milicia africana Sim, For. Fl. Port. E. Afr.: 97, t. 122 (1909). Type: Mozambique, *Sim* 5386 (not seen).
Chlorophora alba A. Chev. in Bull. Soc. Bot. Fr. 58, Mém 8d: 209 (1912). Type from Dahomey.

Tree up to 30(50) m. tall. Leaf lamina elliptic to oblong, 6–20(33) × 3.5–10(12) cm., subcoriaceous, chartaceous when juvenile, apex acuminate, base cordate to obtuse, margin subentire to repand, when juvenile serrate to crenate-dentate; superior surface glabrous or puberulous to pubescent on the main veins, when juvenile often scabridulous; inferior surface densely puberulous on the reticulum, pubescent to puberulous or almost glabrous on the main veins, when juvenile the whole surface hirtellous to tomentose; lateral veins 10–22 pairs, tertiary venation partly perpendicular to the lateral veins; petiole 1–5 cm. long; stipules 0.5–5 cm. long, caducous. Staminate inflorescences: spike 8–20 cm. long, c. 0.5 cm. in diam., peduncle 0.5–2.5 cm. long. Pistillate inflorescences: spike 2–3 cm. long, c. 0.5 cm. in diam. (to 5 × 1.5 cm. in fruit); stigmas up to 7 mm. long. Fruit ellipsoid, 2.5–3 mm. long.

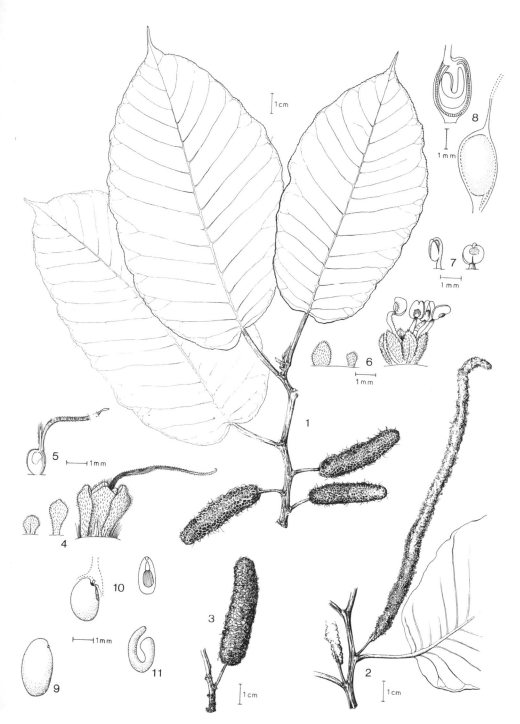

Tab. 6. MILICIA EXCELSA. 1, leafy twig with pistillate inflorescences, *Simao* 152; 2, staminate inflorescence, *Barbosa* 2611; 3, infructescence, *Torre & Paiva* 9372; 4, pistillate flower and bracts; 5, pistil, 4–5 *Leonard* 1049; 6, staminate flower and bracts; 7, stamen, 6–7 *Barbosa* 2278; 8, fruit; 9, endocarp body; 10, seed; 11, embryo, 8–11 *Simao* 654. Drawn by E.H. Hupkens van der Elst and W. Scheepmaker.

Zimbabwe. E: Chipinge Distr., Lusitu R. Valley, below Glencoe For. Res., x.1966, *Goldsmith* 84/66 (K; LISC; PRE; SRGH). S: Chiredzi Distr., Nyahungwe, Runde R., south bank, Gonarezhou Game Reserve, 12.ix.1970, *Sherry* 174/70 (K; LISC; PRE; SRGH). **Malawi**. C: Dedza Distr., Mua-Livulezi Forest, 19.iv.1968, *Salubeni* 1158 (K; MAL; SRGH). **Mozambique**. N: Nampula, on road from Netia to Nacaroa, 27.x.1948, *Barbosa* 2611 (LISC; LMA). Z: Maganja da Costa, 27.x.1942, *Torre* 4710 (LISC). MS: Manica, c. 1 km. de Dombe, 24.x.1953, *Pedro* 4391 (LMA; PRE). GI: Massinga, 3.x.1947, *Pedro & Pedrógão* 2308 (PRE).

From Guinea-Bissau to Ethiopia and southwards to Uganda, Kenya, Tanzania and Angola. Evergreen or semi-deciduous forest; 0–1350 m.

3. MACLURA Nutt.

Maclura Nutt., Gen. N. Amer. Pl. **2**: 233 (1818) nom. conserv. —C.C. Berg in Proc. Konink. Ned. Akad. Weten., ser. C, **89**: 241–247 (1986); in F.T.E.A., Moraceae: 6 (1989).

Chlorophora Gaudich. in Freyc., Voy. Monde, Bot.: 508 (1830). —Benth. & Hook. f., Gen. Pl. **3**, 1: 363 (1880) pro parte excl. sp. Afr.

Cardiogyne Bur. in DC., Prodr. **17**: 232 (1873). —C.C. Berg in Bull. Jard. Bot. Brux. **47**: 359 (1977).

Maclura Nuttall sect. *Cardiogyne* (Bur.) Corner in Gard. Bull., Singapore **19**: 237 (1962).

Shrubs, small trees or climbers, dioecious, armed with spines. Leaves spirally (to subdistichously) arranged, pinnately veined; stipules lateral, free or connate on spine-forming branchlets. Inflorescences in the leaf axils, globose-capitate, bracteate, with yellow dye-containing glands embedded in tepals and bracts; tepals 4, partly connate; stamens 4, inflexed in bud; pistillode present in staminate flower; ovary free; stigmas 1 (or 2 but then unequal in length), filiform. Fruiting perianth enlarged, fleshy, yellow to orange-coloured; fruit free, somewhat drupaceous; seed rather small, without endosperm, cotyledons thin, equal, plicate.

A genus of 11 species in the Old and New Worlds with only one species in Africa.

Maclura pomifera (Rafin.) C.K. Schneider, a native of North America, is cultivated as an ornamental in Zimbabwe. Tree, armed with spines, and with large infructescences up to 10 cm. in diam.

Maclura africana (Bur.) Corner in Gard. Bull., Singapore **19**: 257 (1962). —C.C. Berg in F.T.E.A., Moraceae: 6 (1989). TAB. **7**. Type from Tanzania.

Cardiogyne africana Bur. in DC., Prodr. **17**: 233 (1873). —Kirk in Journ. Linn. Soc., Bot. **9**: 229 (1866) as *"Cudranea"*. —Oliv. in Hook. f., Ic. Pl. **25**, t. 2473 (1896). —Rendle in F.T.A. **6**, 2: 24 (1916). —Hutch. in F.C. **5**, 2: 523 (1920). —F. White, F.F.N.R.: 26 (1962). —K. Coates Palgrave, Trees Southern Africa: 100 (1977). —C.C. Berg in Bull. Jard. Bot. Brux. **47**: 360, t. 22 (1977). Type as above.

Milicia spinosa Sim, For. Fl. Port. E. Afr.: 98, t. 74B (1909). Type: Mozambique, *Sim* 6143 (not seen).

Tangled, spiny shrub, or scrambler up to 7(8) m. tall, or small much-branched tree; branches long, weak; branchlets up to 10 cm. long, ending in a spine. Leaf lamina elliptic to lanceolate (or subcircular), 1.5–9 × 1–4.5 cm., subcoriaceous; apex obtuse to subacute, to shortly acuminate or emarginate; base acute to obtuse; margin entire; superior surface (almost) glabrous, inferior surface sparsely puberulous; lateral veins 4–12 pairs, with the tertiary venation reticulate; petiole 3–30 mm. long; stipules up to 0.2 mm. long, persistent. Staminate inflorescences 0.5–1.5 cm. in diam., peduncle 0.5–2 cm. long. Pistillate inflorescence 0.5–0.8 cm. in diam., to 1.8 cm. in fruit; peduncle 1–5 mm. long; stigmas up to 13 mm. long. Fruit ovoid, 6–7 mm. long.

Zambia. N: Luangwa R., 5.vi.1958, *Fanshawe* 4535 (FHO; K; NDO). C: Luangwa (Feira), 8.ix.1964, *Fanshawe* 8893 (NDO). E: Petauke, Luangwa, N. of Luangwa Bridge, 5.ix.1947, *Brenan & Greenway* 7805 (FHO; K; NDO). **Zimbabwe**. N: Hurungwe (Urungwe), Nemana Pools (?Mana Pools) Area, bank of Zambezi R., 27.ix.1959, *Lovemore* 558 (K; LISC; PRE; SRGH). E: Chipinge Distr., at Mutema Irrigation Scheme, Save (Sabi) Valley, 13.ix.1963, *Plowes* 2340 (PRE; SRGH). S: Mwenezi, near Yangambi R. Drift, c. 8 km. W. of Mateke Hills, 2.v.1958, *Drummond* 5532 (K; LISC; PRE; SRGH). **Malawi**. C: Salima to Balaka road, c. 12 km. S. of Chipoka, 4.v.1980, *Blackmore, Brummitt & Banda* 1441 (K; MAL). S: Nsanje Distr., 5 km. NW. of Nsanje, near Nyamadzere Rest House, 28.v.1970, *Brummitt* 11139 (K; PRE). **Mozambique**. N: Pemba, Maringanha light house, 21.iii.1960, *Gomes e Sousa* 4551 (K; PRE). Z: Entre Marral and Mopeia, 27.vii.1942, *Torre* 4431 (LISC). T: Carinde, 12.iv.1972, *Macêdo* 5191 (PRE). MS: Vila Machado, Lamego, 26.ii.1948, *Mendonça* 3820 (C; LISC; LMA). GI: Massangena, vii.1932, *Smuts* 370 (BM; K; PRE). M: Inhaca I., 23 km. E. of Maputo, xii.1956, *Mogg* 30839 (BM; K; LMA; PRE).

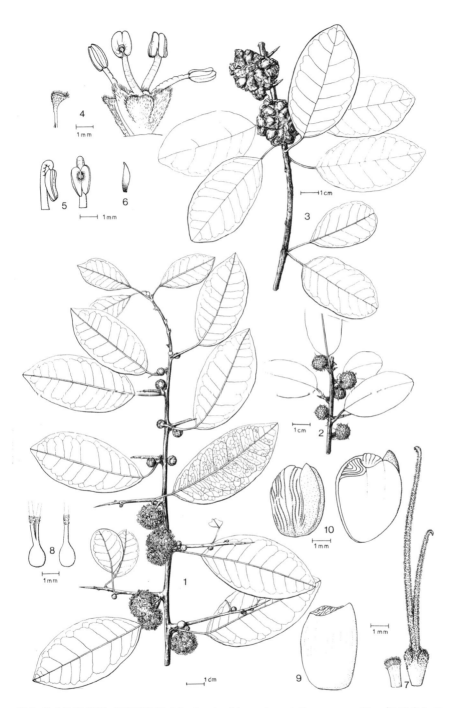

Tab. 7. MACLURA AFRICANA. 1, leafy twig with staminate inflorescences, *Torre* 7142; 2, leafy twig with pistillate inflorescences, *Andrada* 1644; 3, leafy twig with infructescences, *Simao* 603; 4, staminate flower and bracts; 5, stamens; 6, pistillode, 4–6 *Faulkner* 1610; 7, pistillate flower and bract; 8, pistil, 7–8 *Drummond & Hemsley* 2391; 9, seed; 10, embryo, 9–10 *Simao* 603. Drawn by E.H. Hupkens van der Elst and W. Scheepmaker.

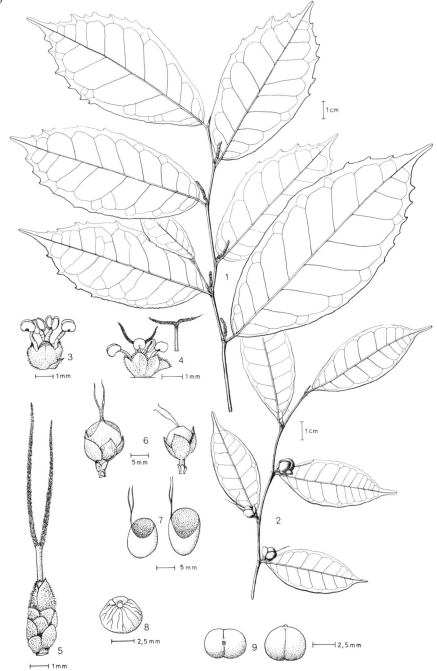

Tab. 8. **STREBLUS USAMBARENSIS.** 1, leafy twig with young staminate inflorescences, *Simao* 336; 2, leafy twig with young infructescences, *Gutzwiller* 2514; 3, staminate flower, *de Wilde* 3542; 4, staminate flower with bract and pistillode, *Hill* 296; 5, pistillate inflorescence, *Breteler* 1865; 6, infructescences; 7, fruits; 8, seed; 9, embryo, 6–9 *Breteler* 1507. Drawn by E.H. Hupkens van der Elst and W. Scheepmaker.

Also from SE. Kenya to South Africa (N. Natal, NE. Transvaal); and from Madagascar. Locally common at low altitudes in riverine vegetation and pan edges, in coastal scrub and on coral rocks; 0–1000 m.

Immersed yellow glands often occur below the thickened part of the tepals and bracts. Normally two glands occur, each beside the midvein, but may be irregularly present, reduced or, in some specimens, lacking. In pistillate flowers they may develop only after the fruit begins to mature.

4. STREBLUS Lour.

Streblus Lour. in Fl. Cochin.: 615 (1790). —C.C. Berg in Proc. Konink. Ned. Akad. Weten., ser. C, **91**: 356 (1988).
Ampalis Boj. in Hort. Maurit.: 291 (1837).
Streblus subgen. *Parastreblus* Blume in Mus. Bot. Lugd.-Bat. **2**: 80 (1856).
Pachytrophe Bur. in DC., Prodr. **17**: 234 (1873).
Ampalis sect. *Pachytrophe* (Bur.) Baill. in Hist. Plantes **6**: 191 (1877).
Sloetiopsis Engl., Bot. Jahrb. **39**: 573 (1907). —C.C. Berg in Bull. Jard. Bot. Brux. **47**: 363 (1977); in F.T.E.A., Moraceae: 8 (1989).
Neosloetiopsis Engl. in Bot. Jahrb. **51**: 426 (1914).

Shrubs or small trees, dioecious or sometimes monoecious. Leaves distichous, pinnately veined; stipules semi-amplexicaul, free. Inflorescences in the leaf axils, bracteate. Staminate inflorescences spicate; tepals 4, basally connate; stamens 4, inflexed in bud, pistillode present. Pistillate inflorescences uniflorous; tepals 4, free; ovary free; stigmas 2, filiform, subequal in length. Fruiting perianth enlarged, hardly fleshy, green; fruit a free dehiscent drupe, the white fleshy exocarp pushing out the black endocarp body (pyrene); seed large, without endosperm, cotyledons thick, equal.

An Old World genus of 34 species, one occurring in Africa and two in Madagascar.

Streblus usambarensis (Engl.) C.C. Berg in Proc. Konink. Ned. Akad. Weten., ser. C, **91**: 357 (1988). TAB. 8. Type from Tanzania.
Sloetiopsis usambarensis Engl., Bot. Jahrb. **39**: 573, tab. on p. 574 (1907). —Rendle in F.T.A. **6**, 2: 77 (1916). —C.C. Berg in Bull. Jard. Bot. Brux. **47**: 364, t. 23 (1977); in Fl. Cameroun **28**: 12, t. 3 (1985); in F.T.E.A., Moraceae: 8 (1989). Type as above.
Neosloetiopsis kamerunensis Engl., Bot. Jahrb. **51**: 426, t. 1 (1914). —Rendle in F.T.A. **6**, 2: 78 (1916). —Hauman in F.C.B. **1**: 82 (1948). —Keay in F.W.T.A. ed. 2, **1**: 595 (1958). Type from Cameroon.

Shrub or small tree, up to 5 m. tall. Leaf lamina oblong to elliptic (lanceolate), (1)2–16 × (0.5)1.5–6 cm., subcoriaceous; apex acuminate to subcaudate, base rounded to acute, margin, at least towards the apex, crenate to serrate-dentate or subentire; both surfaces almost glabrous; lateral veins 4–13 pairs, with the tertiary venation reticulate. Petiole 2–7 mm. long; stipules 2–8 mm. long, often subpersistent. Staminate inflorescences: spike 5–50 mm. long, c. 4 mm. thick, subsessile or with a peduncle up to 1.5 mm. long. Pistillate inflorescences: peduncle 2–3 mm. long, to 5 mm. long in fruit; tepals c. 2 mm. long, to 5 mm. in fruit; stigmas (2)6–8 mm. long. Fruit c. 10 mm. long; endocarp body subglobose, c. 10 mm. in diam., black.

Mozambique. MS: Inhamitanga Forest, Inhamitanga, iv.1970, *Tinley* 1912 (K; PRE; SRGH).
Also from Guinea to E. Zaire, SE. Kenya and NE. Tanzania. Forest and forest edges and along streams; 0–1000 m.

5. TRECULIA Decne.

Treculia Decne. in Ann. Sci. Nat., sér. 3, **8**: 108 (1847). —C.C. Berg in Bull. Jard. Bot. Brux. **47**: 378 (1977); in F.T.E.A., Moraceae: 8 (1989).

Trees, dioecious or sometimes monoecious. Leaves almost distichous, pinnately veined; stipules fully amplexicaul, free. Inflorescences unisexual, sometimes bisexual, borne in the leaf axils and/or (especially the pistillate ones) on the older wood down to the trunk, globose to obovoid-capitate, with a thick rhachis and numerous peltate, long stipitate bracts. Staminate flowers: perianth 2–4(5)-lobed; stamens 2–4, straight in bud;

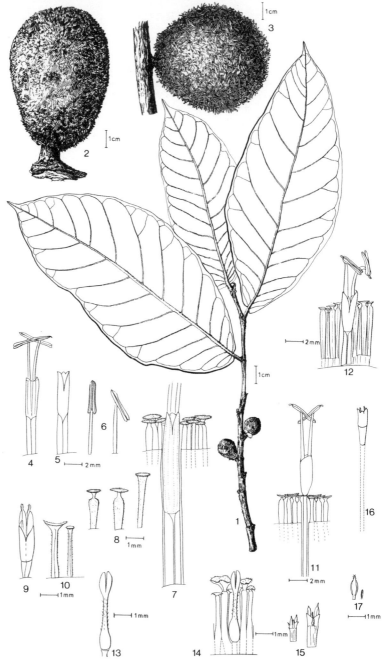

22

Tab. 9. TRECULIA AFRICANA subsp. AFRICANA var. AFRICANA. 1, leafy twig with young inflorescences, *Mpom* 134; 2, staminate inflorescence, *Le Testu* 3848; 3, staminate inflorescence, *de Wilde* 2605; 4, staminate flower; 5, perianth; 6, stamens; 7, staminate flower and bracts; 8, bracts, 4–8 *de Wilde* 2662; 9, staminate flower; 10, bracts, 9–10 *Callens* 2886; 11, staminate flower and bracts, *Le Testu* 3848; 12, staminate flower and bracts, *Zenker* 2525; 13, young pistillate flower; 14, young pistillate flower and bracts; 15, abortive staminate flower, 13–15 *Le Testu* 3831; 16, abortive staminate flower in pistillate inflorescence; 17, stamen and pistillode of abortive staminate flower, 16–17 *Leeuwenberg* 10217. Drawn by E.H. Hupkens van der Elst and W. Scheepmaker.

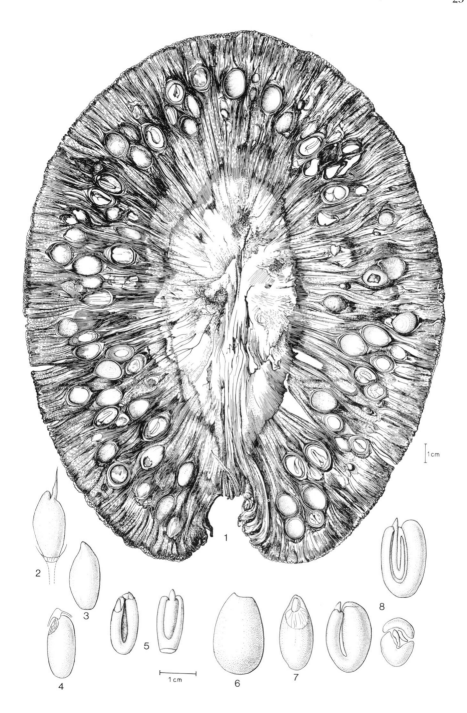

Tab. 10. TRECULIA AFRICANA subsp. AFRICANA var. AFRICANA. 1, infructescence, *Capuron* 6894; 2, fruit; 3, endocarp body; 4, seed; 5, embryo, 2–5 *Leeuwenberg* 10217; 6, endocarp body; 7, seed; 8, embryo, 6–8 *Espirito Santo* 5. Drawn by E.H. Hupkens van der Elst and W. Scheepmaker.

pistillode usually absent. Pistillate flowers without a perianth, stigmas 2, filiform, equal. Fruits somewhat drupaceous, embedded in the soft middle layer of the infructescence; seed large, with remnants of endosperm; cotyledons unequal, curved, one thick, the other thin.

A genus of three species in Africa and Madagascar.

Treculia africana Decne. in Ann. Sci. Nat., sér. 3, **8**: 108, t. 3 (1847). TABS. **9**, **10**. Type from "Senegambia".

Subsp. **africana**

Var. **africana** —Hook. f., Bot. Mag. **98**, t. 5986 (1872). —Hutch. in F.T.A. **6**, 2: 227 (1917). —Hauman in F.C.B. **1**: 90 (1948). —Keay in F.W.T.A. ed. 2, **1**: 613 (1958). —C.C. Berg in Bull. Jard. Bot. Brux. **47**: 382, t. 28, 30 (1977); in Fl. Cameroun **28**: 16, t. 4 (1985).
  Treculia affona N.E. Br. in Bull. Misc. Inf., Kew **1894**: 360 (1894). Type from Nigeria.
  Treculia africana var. *nitida* Engl., Mon. Afr. Pflanzen. **1**: 33 (1898). Type from Cameroon.
  Treculia dewevrei De Wild. & T. Durand in Ann. Mus. Congo, Bot., sér. 2, **1**: 54 (1899). Type from Zaire.
  Treculia engleriana De Wild. & T. Durand in Ann. Mus. Congo, Bot., sér. 1, **1**: 140 (1900). Type from Zaire.
  Ficus whytei Stapf in Johnston, Liberia **2**: 650 (1906). Type from Liberia.
  Artocarpus? africanus Sim, For. Fl. Port. E. Afr.: 102, t. 32 (1909). Type: Mozambique, *Sim* 5999 (type not seen).
  Treculia africana var. *engleriana* (De Wild. & T. Durand) Engl., Pflanzenw. Afr. **3**, 1: 30, t. 8 (1915). Type as above.

Tree up to 30(50) m. tall. Leaf lamina oblong-lanceolate, sometimes elliptic or ovate, (5)10–25(50) × (2.5)4–12(20) cm., coriaceous; apex acuminate, sometimes subacute; base cuneate to subcordate; margin entire to faintly repand; superior surface subglabrous, inferior surface sparsely puberulous on the main veins; lateral veins (8)10–18 pairs, with the tertiary venation mainly reticulate; petiole 2–15 mm. long; stipules 1–1.8 cm. long, caducous. Inflorescences globose, ellipsoid or obovoid, 2.5–10 cm. in diam.; peduncle up to 4 mm. long; stigmas 3.5(10) mm. long. Infructescences globose or nearly so, up to 30 cm. in diam.; fruit ellipsoid to oblong, 10–15 mm. long.

Zambia. N: Kawambwa, swamp near Kawambwa Boma, 8.x.1952, *White* 3639 (FHO; PRE). W: Solwezi, 20.xi.1950, *Fanshawe* 5067 (FHO). **Malawi**. N: Nkhota-Kota Distr., Chia area, 1.ix.1946, *Brass* 17464 (K; NY; PRE). S: Thyolo Distr., Kwerside at Mboma, 20.vii.1943, *Hornby* 2899 (PRE). **Mozambique**. N: Malema Distr., Malema R., 28.x.1954, *Gomes e Sousa* 6269 (COI; K; PRE). Z: Mocuba Distr., Namagoa, c. 200 km. inland from Quelimane, ix–xii.1944, *Faulkner* P42 (K; PRE). M: Libombos, 1908, *Sim* 2050 (PRE).
Also from Senegal to southern Sudan and northern Angola. Riverine, mixed evergreen and swamp forests, woodlands; 0–1300 m.

Subsp. *africana* var. *mollis* (Engl.) J. Léon., with a dense tomentum on the leaves, twigs and stipules, is known only from Nigeria, Cameroon, Gabon and Zaire. Subsp. *madagascarica* (N.E. Br.) C.C. Berg has several varieties in Madagascar and is not easily separated as a whole from the continental subspecies.

6. ANTIARIS Leschen.

Antiaris Leschen. in Ann. Mus. Hist. Nat. Paris **16**: 478 (1810) nom. conserv. —C.C. Berg in F.T.E.A., Moraceae: 10 (1989).
Ipo Pers., Syn. Pl. **2**: 566 (1807).

Trees, monoecious or dioecious; lateral branches self-pruning. Leaves distichous on the lateral branches, pinnately veined; stipules semi-amplexicaul, free. Inflorescences on minute spurs, in the leaf axils or just below the leaves, unisexual, involucrate. Staminate inflorescences pedunculate; flowers numerous; tepals 2–7, free; stamens 2–4, straight in bud; pistillode absent. Pistillate inflorescences sessile or pedunculate, uniflorous; perianth partly adnate to the receptacle, 4-lobed; ovary adnate to the perianth, stigmas 2, ligulate, equal. Fruit forming a drupaceous whole together with the enlarged fleshy, orange to scarlet-coloured receptacle; seed large, without endosperm; cotyledons thick, equal.

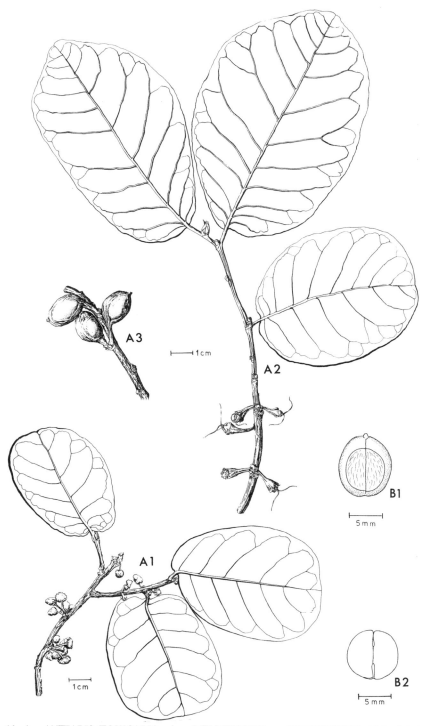

Tab. 11. A.—ANTIARIS TOXICARIA subsp. WELWITSCHII var. WELWITSCHII. A1, leafy twig
with staminate inflorescences, *Flamigni* 10490; A2, leafy twig with pistillate inflorescences,
Devred 762; A3, twig with infructescences, *Toka* 26. B.—ANTIARIS TOXICARIA var.
USAMBARENSIS. B1, seed; B2, cross-section of embryo, B1–2 *Gille* 282.

A monotypic genus confined to the Old World.

Antiaris toxicaria Leschen. in Ann. Mus. Hist. Nat. Paris **16**: 478, t. 22 (1810). —Corner in Gard. Bull., Singapore **19**: 244 (1962). —C.C. Berg in Bull. Jard. Bot. Brux. **47**: 309, 310 (1977); in Fl. Cameroun **28**: 106, t. 36, 37 (1985); in F.T.E.A., Moraceae: 10 (1989). Type from Java.

Tree up to 40(60) m. tall. Leaf lamina elliptic to oblong or almost obovate, when juvenile often lanceolate, (2)6–15(32) × (1.5)3–12 cm., coriaceous, often chartaceous when juvenile; apex shortly acuminate to obtuse or subacute; base obtuse to subcordate or sometimes subacute; margin subentire or denticulate, often dentate when juvenile; superior surface puberulous or scabridulous, hirtellous on the midrib; inferior surface puberulous, hispidulous, sometimes tomentose; lateral veins (5)7–14 pairs, tertiary venation partly scalariform; petiole 3–10 mm. long; stipules 3–10(15) mm. long, caducous. Staminate inflorescences 0.6–1.2(2) cm. in diam.; peduncle 5–15(18) mm. long. Pistillate inflorescences 3–4 cm. in diam., sessile or with a peduncle 3–6 mm. long; stigmas (2)5–8(10) mm. long. Infructescences ellipsoid, sometimes ovoid or globose, 1–1.5 × 0.8–1 cm.

Subsp. **welwitschii** (Engl.) C.C. Berg in Bull. Jard. Bot. Brux. **48**: 466 (1978); in Fl. Cameroun **28**: 106, t. 36 (1985). Lectotype from Angola, chosen by Berg et al. in Fl. Cameroun **28**: 108 (1985).

Var. **welwitschii** (Engl.) Corner in Gard. Bull., Singapore **19**: 248 (1962). —C.C. Berg in Bull. Jard. Bot. Brux. **48**: 466 (1978); in F.T.E.A., Moraceae: 13 (1989). TAB. **11**. Type as above.
    Antiaris welwitschii Engl., Bot. Jahrb. **33**: 118 (1902). —Hutch. in F.T.A. **6**, 2: 224 (1917). —Hauman in F.C.B. **1**: 93, photograph 6 (1948). —Keay in F.W.T.A. ed. 2, **1**: 613 (1958). Type as above.
    Antiaris toxicaria subsp. *africana* var. *welwitschii* (Engl.) C.C. Berg in Bull. Jard. Bot. Brux. **47**: 316, t. 10, fig. 1–3 (1977). Type as above.

Leaf lamina subcoriaceous, margin subentire; superior surface smooth, the midrib puberulous to hirtellous, inferior surface smooth or the midrib sparsely appressed-puberulous, occasionally hirtellous; only the midrib and lateral veins prominent beneath, the smaller veins plane or almost so.

Zambia. N: Samfya, L. Bangweulu, 18.xi.1964, *Mutimushi* 1167 (K; NDO).
From Sierra Leone eastwards to northern Tanzania and southwards to Angola. Evergreen swamp forest, or riverine forest (or semi-deciduous forest); 0–1300 m.

Var. **usambarensis** (Engl.) C.C. Berg in Bull. Jard. Bot. Brux. **48**: 468 (1978). TAB. **11**. Type from Tanzania.
    Antiaris usambarensis Engl., Bot. Jahrb. **33**: 119 (1902). —Hutch. in F.T.A. **6**, 2: 224 (1917). —Hauman in F.C.B. **1**: 94 (1948). Type as above.
    Antiaris toxicaria subsp. *africana* var. *usambarensis* (Engl.) C.C. Berg in Bull. Jard. Bot. Brux. **47**: 309, 318, t. 10 (1977). Type as above.

Leaf lamina subcoriaceous, margin subentire; superior surface smooth to scabridulous, puberulous at least on the midrib, or glabrous; inferior surface scabridulous to scabrous, hirtellous or hispidulous to puberulous on the veins; the midrib, lateral veins and part of the smaller veins prominent beneath, the reticulum almost plane.

Zambia. N: Mbala Distr., Lunzua R., 21.i.1960, *Lawton* 678 (K).
Also in Uganda, Kenya and Tanzania. Riverine and evergreen forest; 1000–1800 m.

A. toxicaria subsp. *welwitschii* var. *africana* occurs from Senegal to southern Sudan, Uganda and Zaire. The varieties of subsp. *welwitschii* are not very clear-cut morphologically. Juvenile specimens are more similar and can often hardly be told apart.

7. BOSQUEIOPSIS De Wild. & T. Durand

Bosqueiopsis De Wild. & T. Durand in Bull. Herb. Boiss., sér. 2, **1**: 839 (1901). —C.C. Berg in Bull. Jard. Bot. Brux. **47**: 293 (1977); in F.T.E.A., Moraceae: 15 (1989).

Trees or shrubs, monoecious (?) sometimes androdioecious. Leaves distichous, pinnately veined to subtriplinerved; stipules fully amplexicaul, free. Inflorescences in the leaf axils or just below the leaves, bisexual or staminate, discoid or subglobose to

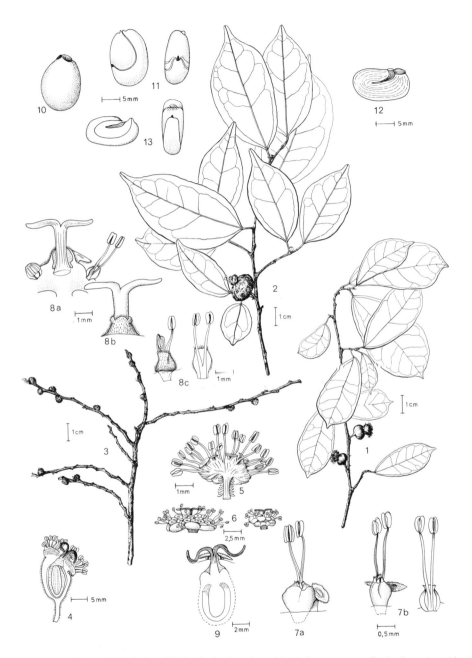

Tab. 12. BOSQUEIOPSIS GILLETII. 1, leafy twig with inflorescences; 2, leafy twig with infructescence, 1–2 *Carlier* 272; 3, leafless twig with staminate inflorescences, *Schlieben* 5437; 4, bisexual inflorescence; 5, staminate inflorescence, 4–5 *Gillet* 17956; 6, staminate inflorescences, *Collin* 17; 7a–b, staminate flower, *Gillet* 17956; 8a–c, abortive pistillate flower and staminate flowers, *Collin* 17; 9, pistillate flower, *Gillet* 17956; 10, seed; 11, embryo, 10–11 *Torre & Paiva* 10051; 12, seed; 13, embryo, 12–13 *Anonymous* s.n. Drawn by E.H. Hupkens van der Elst and W. Scheepmaker.

turbinate-capitate; bracts interfloral, peltate, marginal bracts often basally attached. Staminate flowers several to many; tepals 3–4, connate; stamens (1)2, inflexed in bud; pistillode present. Pistillate flower 1, central, partly adnate to the receptacle; perianth 4-lobed; ovary adnate to the perianth; stigmas 2, band-shaped, equal. Fruit forming a drupaceous whole with an enlarged, fleshy, orange to yellow-coloured receptacle, crowned with the remnants of staminate flowers and bracts; seed large, without endosperm; cotyledons thick, unequal.

A monotypic, tropical African genus.

Bosqueiopsis gilletii De Wild. & T. Durand in Bull. Herb. Boiss., sér. 2, **1**: 840 (1901). —Engl., Bot. Jahrb. **51**: 435, t. 2 (1914). —Hutch. in F.T.A. **6**, 2: 217 (1917). —Hauman in F.C.B. **1**: 96 (1948). —C.C. Berg in Bull. Jard. Bot. Brux. **47**: 294, t. 5 (1977); in F.T.E.A., Moraceae: 17 (1989). TAB. **12**. Type from Zaire.
    Bosqueiopsis lujae De Wild., Pl. Nov. Herb. Hort. Then. **7**: 239, t. 56 (1977). Type from Zaire.
    Bosqueiopsis carvalhoana Engl., Bot. Jahrb. **51**: 436, t. 3 (1914). Type: Mozambique, without precise locality, *Carvalho* s.n. (COI, holotype; B, isotype).
    Bosqueiopsis parvifolia Engl., Bot. Jahrb. **51**: 437, t. 4 (1914). —Hutch. in F.T.A. **6**, 2: 216 (1917). Type from Tanzania.
    Trymatococcus parvifolius Engl., Pflanzenw. Afr. **3**, 1: 27, t. 15 (1915), without description, drawing based on type of *B. parvifolia*.

Shrub to 6 m. tall or tree up to 35 m. tall. Leaf lamina oblong to ± broadly oblanceolate, sometimes elliptic, 2–14(21) × 1.5–6(12.5) cm., subcoriaceous to chartaceous; apex acuminate, base ± cuneate, margin entire; superior surface glabrous, inferior surface puberulous; lateral veins 5–6 pairs, with the tertiary venation reticulate; petiole 3–10(15) mm. long; stipules 2–7 mm. long, caducous. Bisexual inflorescences 5–10 mm. in diam., subsessile or with a peduncle up to 4 mm. long; stigmas 2–4.5 mm. long. Staminate inflorescences 3–6 mm. in diam., subsessile or peduncle up to 2 mm. long. Infructescences subglobose to ellipsoid, c. 2 cm. in diam.

Mozambique. N: 2 km. de Itoculo para o Régulo Chihir, 2.xii.1963, *Torre & Paiva* 9363 (PRE). Also in Congo, Zaire and eastern Tanzania. Deciduous thicket (Tanzania) and open forest.

In the Congo and Zaire this species forms trees to 35 m. tall, whereas in the east African coastal thickets the shrubs or treelets only reach 6 m. or so. These may represent a subspecies.

8. TRILEPISIUM Thouars

Trilepisium Thouars, Gen. Nov. Madag.: 22 (1806). —DC., Prodr. **2**: 639 (1825). —C.C. Berg in Bull. Jard. Bot. Brux. **47**: 297 (1977); in F.T.E.A., Moraceae: 17 (1989).
    Bosqueia Baill. in Adansonia **3**: 338 (1863).
    Pontya A. Chev. in Bull. Soc. Bot. Fr. **58**, Mém. 8d: 210 (1912).

Trees, monoecious. Leaves distichous, pinnately veined (to subtriplinerved); stipules fully amplexicaul, connate. Inflorescences in the leaf axils, bisexual, initially enveloped by 2 coriaceous bud-scales, involucrate. Staminate flowers peripheral, without a perianth, initially covered by the membranous expanded margin of the receptacle, (this cover tearing at anthesis, leaving a tubular part around the pistillate flower and a fringe-like marginal part); stamens without distinct floral arrangement, straight before anthesis. Pistillate flower 1, central, partly adnate to the receptacle; perianth 4-lobed; ovary adnate to the perianth, stigmas 2, ligulate, equal. Fruit forming a drupaceous whole with the enlarged, fleshy, dark purple to red receptacles crowned with the remnants of the stamens and the marginal part of the receptacle. Seed large, without endosperm; cotyledons equal, thick and fused.

A monotypic, tropical African genus.

Trilepisium madagascariense DC., Prodr. **2**: 639 (1825). —C.C. Berg in Bull. Jard. Bot. Brux. **47**: 299, t.6, 7 (1977); in Fl. Cameroun **28**: 103, t. 35 (1985); in F.T.E.A., Moraceae: 17 (1989). TAB. **13**. Type from Madagascar.
    Bosqueia thouarsiana Baill. in Adansonia **3**: 339, t. 10 (1863). Type from Madagascar.
    Bosqueia thouarsiana var. *acuminata* Baill. in Adansonia **3**: 339 (1863). Type from Madagascar.

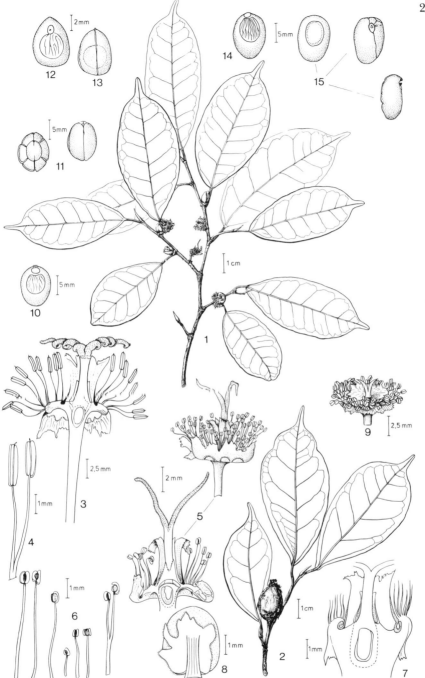

Tab. 13. TRILEPISIUM MADAGASCARIENSE. 1, leafy twig with inflorescences; 2, leafy twig with infructescence, 1–2 *Breteler* 2732; 3, inflorescence; 4, stamens, 3–4 *Gossweiler* 4394; 5, inflorescence; 6, stamens, 5–6 *Simao* 506; 7, inflorescence; 8, involucre, 7–8 *Gossweiler* 6514; 9, inflorescence, *Mendonça* 141; 10, seed; 11, embryo, 10–11 *Breteler* 2732; 12, seed; 13, embryo, 12–13 *Capuron* 948; 14, seed; 15, embryo, 14–15 *Devred* 2454. Drawn by E.H. Hupkens van der Elst and W. Scheepmaker.

Bosqueia thouarsiana var. *pyriformis* Baill. in Adansonia **3**: 339 (1863). Type from Madagascar.
Bosqueia boiviniana Baill. in Adansonia **3**: 340, t. 10 (1863). Type from Madagascar.
Bosqueia phoberos Baill. in Adansonia **3**: 339 (1863); op. cit. **8**: 72, t. 4 (1867). —Hutch. in F.T.A.
6, 2: 219 (1917). —K. Coates Palgrave, Trees Southern Africa: 102 (1977). Type from Tanzania.
Bosqueia gymnandra Bak., Fl. Maur. Seych.: 283 (1877). Type from the Seychelles.
Bosqueia angolensis Ficalho, Pl. Ut. Afr. Port.: 271 (1884). —Hutch. in F.T.A. **6**, 2: 218 (1917).
—Hauman in F.C.B. **1**: 95 (1948). —Keay in F.W.T.A. ed. 2, **1**, 2: 612 (1958). Type from Angola.
Bosqueia welwitschii Engl., Mon. Afr. Pflanzen. **1**: 36 (1898). Type from Angola.
Bosqueia cerasifolia Engl., Mon. Afr. Pflanzen. **1**: 36 (1898); Bot. Jahrb. **51**: 439, fig. 5 (1914).
Type from Tanzania.
Pontya excelsa A. Chev. in Bull. Soc. Bot. Fr. **58**, Mém. 8d: 210 (1912). Lectotype from Guinea,
chosen by Berg et al. in Fl. Cameroun **28**: 103 (1985).
Bosqueia danguyana Léandri in Notul. Syst. **13**: 178 (1948). Type from Madagascar.
Bosqueia calcicola Léandri in Notul. Syst. **13**: 179 (1948). Type from Madagascar.
Bosqueia orientalis Léandri in Notul. Syst. **13**: 180 (1948). Type from Madagascar.
Bosqueia manongarivensis Léandri in Notul. Syst. **13**: 180 (1948). Type from Madagascar.
Bosqueia occidentalis Léandri in Notul. Syst. **13**: 181 (1948). Type from Madagascar.

Tree up to 25(40) m. tall. Leaf lamina elliptic to oblong or subobovate to oblanceolate, 2–12(18) × 1.5–6.5(8) cm., subcoriaceous; apex acuminate, base ± broadly cuneate, margin entire; both surfaces glabrous; lateral veins 4–10(12) pairs, with the tertiary venation reticulate; petiole 3–15 mm. long; stipules 2–12 mm. long, glabrous, caducous. Inflorescences 0.5–0.8(1) cm. in diam.; peduncle 0.2–1.5 cm. long, increasing to 2.3 cm. long in fruit. Anthers 0.2–0.8 mm. long, filaments 2–10 mm. long. Stigmas 2–8 mm. long. Infructescence ovoid to ellipsoid, 1.2–1.8 cm. in diam.

Zambia. N: Mbala Distr., Lunzua, 2.iv.1960, *Fanshawe* 5628 (FHO; NDO). **Zimbabwe**. C: Wedza Mt., 19.ii.1963, *Wild & Drummond* 6000 (K; SRGH). E: Chipinge Distr., Chirinda Forest, ix.1967, *Goldsmith* 86/67 (COI; K; LISC; PRE; SRGH). **Malawi**. N: Viphya, Kavuma stream, Kawendama, 10.x.1964, *Chapman* 2293 (FHO). C: Ntchisi Mt., 5.ix.1929, *Burtt Davy* 21258 (FHO). S: Thyolo Distr., Thyolo Mt., 21.ix.1946, *Brass* 17714 (K; PRE). **Mozambique**. N: c. 40 km. de Malema (Entre Rios) estrada para Ribáuè, Serra Murripa, 15.xii.1967, *Torre & Correia* 16536 (LISC). Z: Milange, sul do Monte Tumbine, 30.vii.1949, *Andrada* 1803 (LISC). MS: Manica, Moribane to Sanguene, 6.x.1953, *Pedro* 4229 (K; PRE).

From Guinea to S. Ethiopia, E. Africa and southwards to South Africa and Angola. Also from Madagascar and the Seychelles. Evergreen or semi-deciduous forest; 0–2000 (2500?) m.

9. DORSTENIA L.

By M.E.E. Hijman

Dorstenia L., Sp. Pl. **1**: 121 (1753); Gen. Pl., ed. 5: 56 (1754). —Hijman in F.T.E.A.,
Moraceae: 20 (1989).
Kosaria Forssk., Fl. Aegypt.-Arab.: 164 (1775).
Ctenocladus Engl., Bot. Jahrb. **57**: 246 (1921).
Craterogyne Lanjouw in Rec. Trav. Bot. Néerl. **32**: 272 (1935).
Ctenocladium Airy Shaw in Kew Bull. **18**: 272 (1965).

Herbs, often at least somewhat succulent, rhizomatous or tuberous, or shrubs. Leaves spirally or distichously arranged, pinnately veined, less often palmately or radiately veined; stipules lateral, free. Inflorescences bisexual, borne in the leaf axils, discoid to turbinate, or naviculate, circular, elliptic or stellate in outline, receptacle mostly with marginal and/or submarginal appendages, sometimes also bracteate outside; flowers connate. Staminate flowers numerous; tepals (1)2–3(4), free or basally connate; stamens 2–3; pistillode usually absent. Pistillate flowers (in the central part of the inflorescence) 1–numerous; perianth tubular with only the apex free from the surrounding flowers; ovary free; stigmas 2, filiform to ligulate, equal or unequal in length, or 1. Fruit a dehiscent drupe; the white fleshy part pushing out the endocarp body (pyrene) if large, or ejecting the endocarp body if small; endocarp body (especially if large) subglobose and smooth or tetrahedral and tuberculate. Seed large and without endosperm, or small and with endosperm; cotyledons thick and unequal or flat and equal.

A genus of more than 100 species, with c. 45 species in the Neotropics, 1 in Asia and c. 57 in Africa.

1. Plants with petioles and peduncles arising directly from the apices of branches of a creeping rhizome; leaves ± rosulate; internodes remaining short - - - - 4. *zambesiaca*
 - Plants with a distinct, erect or creeping, aerial stem; leaves well spaced on the stem with internodes more than 5 mm. long - - - - - - - - - - 2
2. Appendages in 1 row from the margin of the receptacle, 2–many, variously triangular to linear or filiform - - - - - - - - - - - - - - - 3
 - Appendages in 2 rows; the inner (marginal) row consisting of distinct filiform or subulate appendages or reduced to a crenate or subentire rim; the outer row, inserted below the receptacle margin, consisting of 2–many triangular to filiform or spathulate appendages - - - - - - - - - - - - - - - 7
3. Face of receptacle ± isodiametric with appendages of varying length - - - 4
 - Face of the receptacle elongate, with ± well developed appendages at either end, with or without shorter lateral appendages - - - - - - - - - - - - 5
4. Margin of the receptacle 3–5 mm. wide with conspicuous radiating stripes, and many tooth-shaped appendages to 4 mm. long, terminal ones to 7 mm. long - - 2. *schliebenii*
 - Margin of the receptacle 0–1.5 mm. wide, without stripes; terminal appendages 7–45 mm. long - - - - - - - - - - - - - - 1. *tayloriana*
5. Lateral appendages absent - - - - - - - - - - 3. *psilurus*
 - Lateral appendages present - - - - - - - - - - - 6
6. Margin of the receptacle 3–5 mm. wide, with conspicuous radiating stripes, and many tooth-shaped appendages to 4 mm. long, terminal ones to 7 mm. long - - 2. *schliebenii*
 - Margin of the receptacle 0–1.5 mm. wide, without stripes; terminal appendages 7–45 mm. long - - - - - - - - - - - - - 1. *tayloriana*
7. Receptacle elongate, boat-shaped with 2 long outer appendages, the longer one generally 5–13 cm. long - - - - - - - - - - - - - 6. *buchananii*
 - Receptacle elliptic to isodiametric, normally with 3–many outer appendages - - 8
8. Stems erect or ascending, sappy or succulent, becoming swollen and sometimes pear-shaped or globose toward the base, without scale leaves; plants rhizomatous; leaves glabrous or sparsely puberulous mostly on the veins, usually repand or coarsely to irregularly toothed, or entire - - - - - - - - - - - 5. *hildebrandtii*
 - Stems annual, erect, herbaceous, from a discoid to subglobular tuber or series of superposed tubers, with scale leaves on lower part; leaves mostly hairy on both surfaces, generally entire to finely denticulate, sometimes coarsely dentate - - - - - - - 9
9. Petiole 0–2(5) mm. long; lamina usually ± scabrous; receptacle variously shaped but not stellate, stigmas (1)2 - - - - - - - - - - 7. *benguellensis*
 - Petiole (2)5–25 mm. long; leaves usually smooth; receptacle stellate or substellate; stigma 1 - - - - - - - - - - - 8. *cuspidata*

1. **Dorstenia tayloriana** Rendle in Journ. Bot. **53**: 300 (1915); in F.T.A. **6**, 2: 43 (1916). —Hijman in F.T.E.A., Moraceae: 25 (1989). Type from Kenya.

Herb up to 50 cm. tall, rhizomatous; stems ascending, unbranched, leafy in the upper part, 1.5–4 mm. thick, whitish- sometimes yellowish-puberulous or hirtellous, the lower part woody. Leaves spirally arranged; lamina oblanceolate to obovate, rarely linear, (1)3–14 × (0.5)1.6 cm., chartaceous when dry, apex acute to obtuse, base obtuse to subcordate or cuneate, margin repand to faintly or sometimes coarsely dentate, both surfaces smooth, or sometimes scabrous, inferior surface puberulous to hirtellous on the main veins; lateral veins 3–8(10) pairs; petiole 2–13 mm. long, 0.5–1.5 mm. thick; stipules subulate, 4–10 mm. long and persistent, or stipules triangular c. 0.1 mm. long and caducous. Inflorescences solitary; peduncle 0.6–2.7 cm. long, gradually passing into the receptacle. Receptacle zygomorphic to almost actinomorphic, naviculate to broadly funnel-shaped and orbicular in outline, flowering face elliptic or subcircular, 0.5–2.2 × 0.4–1 cm., plane, often dark purplish, margin 0.1–1.5 mm. wide, often with a ridge, bordering the flowering face; appendages 10–35, lateral, triangular to linear, (0.2)1–17 mm. long, sometimes with terminal, linear appendages to 45 mm. long. Staminate flowers rather spaced; perianth lobes 2–3; stamens 2–3. Pistillate flowers up to c. 10; perianth tubular; stigmas 2. Endocarp body tetrahedral in shape, c. 2.5 mm. in diam., two sides tuberculate.

Var. **laikipiensis** (Rendle) Hijman in F.T.E.A., Moraceae: 26 (1989). Type from Kenya.
 Dorstenia laikipiensis Rendle in Journ. Bot. **53**: 299 (1915); in F.T.A. **6**, 2: 34 (1916). Type as above.
 Dorstenia pectinata Peter in Fedde, Repert., Beih. **40**, 2: 74, Descr.: 5, t. 6, fig. 1 (1932). Type from Tanzania.
 Dorstenia rugosa Peter in Fedde, Repert., Beih. **40**, 2: 74, Descr.: 5, t. 5, fig. 2 (1932). Type from Tanzania.

Herb up to 30 cm. tall. Both surfaces of the lamina sometimes scabrous; stipules subulate, 4–10 mm. long, persistent. Receptacle tending to be actinomorphic, broadly naviculate to funnel-shaped, terminal appendages hardly longer than the secondary appendages; flowering face elliptic to subcircular.

Mozambique. Z: Gúruè, Rio Namuka, fl. 31.vii.1979, *Schäfer* 6900 (K).
Also in Tanzania and Kenya. In evergreen and riverine forests.
Var. *tayloriana* occurs in northeast Tanzania and southeast Kenya. It differs mainly in having caducous stipules c. 1 mm. long and a zygomorphic receptacle with terminal appendages distinctly longer than the secondary appendages.

2. **Dorstenia schliebenii** Mildbr. in Notizbl. Bot. Gart. Berl. **11**: 396 (1932). —Hijman in F.T.E.A.,
   Moraceae: 28 (1989). TAB. **14**. Type from Tanzania.
   Dorstenia hispida Peter in Fedde, Repert., Beih. **40**, 2: 74, Descr.: 4, t. 4 (1932) non Hook. (1840)
   nom. illegit. Type from Tanzania.
   Dorstenia kyimbilaensis De Wild., Pl. Bequaert. **6**: 39 (1932). Type from Tanzania.

Herb up to 1 m. tall, rhizomatous; stems erect or ascending, unbranched, up to c. 5 mm. thick, minutely puberulous to yellowish-hirtellous, the lower part woody. Leaves spirally arranged; lamina subobovate to obovate, 6–15 × 2–5 cm., chartaceous to subcoriaceous when dry, apex acute to obtuse, base cuneate to obtuse or cordate, margin entire, sometimes coarsely dentate; superior surface scabrous, the inferior scabrous and partly puberulous, (mainly so on the veins) or minutely puberulous to yellowish-hirtellous; venation impressed and conspicuous above, lateral veins 6–10 pairs; petiole 5–15 mm. long, 1–3 mm. thick; stipules narrowly triangular to subulate, 2–4.5 mm. long, usually persistent. Inflorescences solitary or sometimes in pairs; peduncle 5–12 mm. long; receptacle ± zygomorphic, discoid and plane, elliptic to subcircular or multiangular in outline, 2–3 × 1.2–1.5 cm.; margin 3–5 mm. wide, with conspicuous radiating stripes; appendages in 1 row, primary (terminal) appendages 2, linear to spathulate, 5–7 mm. long; secondary (lateral) appendages numerous, tooth-shaped, up to 4 mm. long. Staminate flowers ± crowded; perianth lobes 2–3. Pistillate flowers less numerous; perianth tubular; stigmas 2. Endocarp body tetrahedral, c. 2.5 mm. in diam., tuberculate.

Malawi. N: c. 8 km. E. of Mzuzu, on road to Nkhata Bay, Rose Falls, fl. 7.iii.1977, *Grosvenor & Renz* 1063 (K; PRE; SRGH).
Also in Tanzania. In riverine and evergreen forests, often in rocky places; up to 2000 m.

3. **Dorstenia psilurus** Welw. in Trans. Linn. Soc. **27**: 71 (1869). —Engl., Mon. Afr. Pflanzen. **1**: 20
   (1898). —Rendle in F.T.A. **6**, 2: 50 (1916). —Hauman in F.C.B. **1**: 69 (1948). —Friis in Norw.
   Journ. Bot. **21**: 101 (1974). —Hijman in Fl. Cameroun **28**: 74, t. 26 (1985); in F.T.E.A., Moraceae:
   31 (1989). Type from Angola.
   Dorstenia bicornis Schweinf. in Bot. Zeit. **29**: 332 (1871); in Bull. Mus. Hist. Nat. Paris Sér. 1, **1**:
   62 (1895) as "*bicuspis*". —Rendle in F.T.A. **6**, 2: 49 (1916). Type from Sudan.
   Dorstenia lukafuensis De Wild. in Ann. Mus. Congo. Bot., Ser. 4, **1**: 28 (1902). Type from Zaire.
   Dorstenia psilurus var. *compacta* De Wild., Pl. Nov. Herb. Hort. Then. **1**: 233 (1907). Type:
   Mozambique, Morrumbala, *Luja* 372 (BR, holotype).
   Dorstenia psiluroides forma *subintegra* Engl. in Mildbr., Wiss. Ergebn. Deutsch. Zentr.-Afr.
   Exped. 1907–1908, **2**: 181 (1911). Type from Zaire.
   Dorstenia psiluroides Engl. in Mildbr., Wiss. Ergebn. Deutsch. Zentr.-Afr. Exped. 1907–1908,
   2: 181 (1911). Type from Zaire.
   Dorstenia stolzii Engl., Bot. Jahrb. **51**: 432 (1914). Type from Tanzania.
   Dorstenia psilurus var. *brevicaudata* Rendle in Journ. Bot. **53**: 301 (1915); in F.T.A. **6**, 2: 51
   (1916). Type from Uganda.

Herb up to 60 cm. tall, rhizomatous, the rhizomes often tuberous; stems erect or ascending, often branched, leafy in the upper part, c. 3(5) mm. thick, puberulous. Leaves spirally arranged, lamina elliptic to obovate, (2)5–19 × (1)2–8 cm., papyraceous when dry, apex acute to acuminate, base cuneate, margin denticulate to coarsely dentate or sometimes repand; both surfaces puberulous; lateral veins 5–8(10) pairs; petiole 5–35 mm. long, 1–2 mm. thick; stipules narrowly triangular, 0.5–2(4) mm. long, subpersistent. Inflorescences solitary or in pairs; peduncle 0.6–5 cm. long, c. 0.5 mm. thick; receptacle naviculate, borne vertically; flowering face narrowly ovate to linear, 1.2–5.5 × 0.15–0.5 cm.; margin almost lacking, sometimes up to 1.5 mm. wide; appendages 2, terminal, filiform, the upper one (20)30–70(100) mm. long, the lower one (1.2)3–10(30) mm. long. Staminate flowers ± spaced; perianth lobes 1–2(3); stamens 1–2(3). Pistillate flowers 5–10(22), most of them in a median row; perianth tubular; stigmas 2. Endocarp body

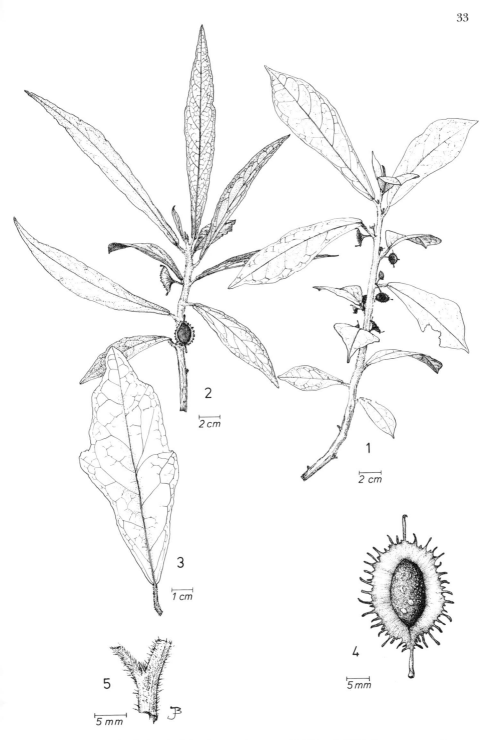

Tab. 14. DORSTENIA SCHLIEBENII. 1, stem, *Drummond & Hemsely* 1753; 2, upper part of stem, *Harris et al.* 5090; 3, leaf; 4, inflorescence; 5, indumentum of stem and stipule, 3–5 *Pawek* 7630. Drawn by J. Brinkman.

subglobose to tetrahedral, c. 3 mm. in diam., at least slightly tuberculate.

Zambia. N: Chinsali Distr., Mbereshi, fl. 16.i.1960, *Richards* 12376 (EA; K; MO; SRGH). W: Mwinilunga Distr., S. of Matonchi Farm, fl. 16.xi.1938, *Milne-Redhead* 3273 (BM; K; PRE). E: Mwangazi R., between Msoro and Great East Road, fl. 6.i.1959, *Robson* 1047 (BM; K; LISC; PRE). **Zimbabwe**. E. Chipinge Distr., Chirinda Forest, path to Big Tree, fl. 20.iii.1970, *Kelly* 176 (K; LISC; PRE; SRGH). **Malawi**. N: Misuku Distr., Misuku Hills, Wilindi Forest, fl. 12.i.1959, *Richards* 10626 (EA; K). C: Dedza Distr., Mua Mission, 21.i.1959, *Jackson* 2310 (BR; K; SRGH). S: Mulanje Mt., Ruo Gorge by Surani Stream, fl. 26.i.1967, *Hilliard & Burtt* 4640 (K). **Mozambique**. N: Meconta, 19 km. from Corrane to Liupo, fl. 18.i.1964, *Torre & Paiva* 10059 (LISC; LISJC). Z: Mocuba Distr., Mugeba, c. 80 km. from Namagoa on road to Pebane, fl. 29.xii.1947, *Faulkner* 130B (K). MS: c. 3 km. from Haroni-Lusitu river confluence, upstream from Lusitu, fl. 23.xi.1967, *Ngoni* 60 (K; LISC; PRE; SRGH).

Also in Tanzania, Uganda, Angola and Cameroon. Evergreen forest floor in shade, riverine forests and high rainfall miombo woodland; up to 1500 m.

Var. *scabra* Bur., from the forests of W. and central Africa, forms clumps with mostly unbranched stems up to 2(3) m. tall.

4. **Dorstenia zambesiaca** Hijman in Kew Bull. **45**: 367 (1990). Type: Mozambique, between Inhamitanga and Lacerdonia on the Zambezi R., *Pope & Müller* 520 (K, holotype; LISC; SRGH, isotype).

Herb with a creeping rhizome; aerial stem 0–15 mm. tall, c. 5 mm. thick, with very short internodes, subglabrous. Leaves spirally arranged, subrosulate; lamina suborbicular to ovate, 3–5 × 3–5.5 cm., papyraceous, apex rounded, base cordate; margin repand to sinuate; both surfaces appressed-puberulous; lateral veins 5–6 pairs; petiole 3–5 cm. long, 1–1.5 mm. thick, appressed-puberulous; stipules subtriangular, c. 2.5 mm. long, ± persistent. Inflorescences solitary, 3–5 mm. in diam.; peduncle up to c. 4 cm. long, c. 1 mm. thick; receptacle broadly funnel-shaped, circular to substellate in outline, 2–5 mm. in diam., appendages in one row; primary appendages 3–5, spathulate, 7–10 mm. long, secondary appendages linear to subspathulate, 0.1–0.3 mm. long. Staminate flowers rather crowded; perianth lobes 3; stamens 3. Pistillate flowers rather numerous; perianth conical; stigmas 2.

Mozambique. MS: between Inhaminga and Lacerdonia on the Zambezi R., 9.xii.1971, *Pope & Müller* 520 (K; LISC; SRGH).

Not known elsewhere. In leaf litter of mixed evergreen forest.

5. **Dorstenia hildebrandtii** Engl., Bot. Jahrb. **20**: 146 (1894); Mon. Afr. Pflanzen. **1**: 23, t. 6b (1898). —Rendle in F.T.A. **6**, 2: 65 (1916). —Peter in Fedde, Repert., Bcih. **40**, 2: 79 (1932). —Hijman in F.T.E.A., Moraceae: 33 (1989). Type from Kenya.

Herb up to c. 70 cm. tall, rhizomatous or tuberous. Stems ascending to erect, branched (the branches often arrested, with minute leaves) or unbranched, fleshy to sometimes thickly succulent, the basal part often swollen, sometimes becoming globose to pear-shaped, up to 4 cm. across, the leafy part 1–5(7) mm. in diam., glabrous to sparsely puberulous; scars of leaves, stipules and inflorescences often conspicuous and prominent. Leaves spirally arranged; lamina oblong to elliptic, lanceolate, oblanceolate or linear, (0.5)1–12(20) × (0.3)1–16 cm., chartaceous to papyraceous when dry, apex obtuse to acute, base cuneate to attenuate to subobtuse; margin irregularly coarsely crenate to dentate, sometimes subentire, ± revolute when dry; superior surface glabrous and smooth, scabridulous or sparsely puberulous on the midrib, the inferior surface puberulous to glabrous; lateral veins 2–10(12) pairs; petiole (0.1)0.2–2(3) cm. long, 0.5–1 mm. thick, puberulous, mainly adaxially; stipules subulate to ovate, up to 1.5(2) mm. long, persistent. Inflorescences solitary (or in pairs); peduncle 2–12(25) mm. long, c. 0.5 mm. thick, sometimes recurved; receptacle discoid to broadly turbinate or funnel-shaped, sometimes subnaviculate, 0.4–1.2 cm. in diam. (or wide); flowering face subcircular to substellate or narrowly elliptic to 3–6-angular in outline; margin very narrow or lacking. Receptacle appendages in 2 rows, those of the inner (marginal) row numerous, semi-circular to triangular, up to 2 mm. long, forming a crenate to dentate rim; appendages of the outer (submarginal) row 2–12, narrowly triangular to band-shaped or filiform, (2)5–22(25) mm. long. Staminate flowers ± crowded; perianth lobes (2)3(4); stamens (2)3(4). Pistillate flowers numerous, scattered; perianth tubular; stigmas 2. Endocarp body tetrahedral, 1–1.5 mm. in diam., slightly tuberculate.

Var. **schlechteri** (Engl.) Hijman in F.T.E.A., Moraceae: 34 (1989). Type: Mozambique, Beira, *Schlechter* s.n. (B, holotype).

    Dorstenia schlechteri Engl., Mon. Afr. Pflanzen. **1**: 23, t. 4e (1898). —Rendle in F.T.A. **6**, 2: 61 (1916). —R.E. Fries in Notizbl. Bot. Gart. Berl. **8**: 607 (1924). —Peter in Fedde, Repert., Beih. **40**, 2: 77 (1932). —Hauman in F.C.B. **1**: 71 (1958). Type as above.

    Dorstenia quercifolia R.E. Fries in Ark. Bot. **13**, 1: 8, t. 1 (1913). Type from Zaire.

    Dorstenia denticulata Peter in Fedde, Repert., Beih. **40**, 2: 74, Descr.: 3, t. 3, fig. 1 (1932). Lectotype from Tanzania, chosen by Hijman in F.T.E.A., Moraceae: 34 (1989).

    Dorstenia polyactis Peter in Fedde, Repert., Beih. **40**, 2: 79, Descr.: 8, t. 9, fig. 1 (1932). Lectotype from Tanzania, chosen by Hijman in F.T.E.A., Moraceae: 34 (1989).

Herb up to 70 cm. tall; stems slender, the base not swollen nor tuber-like, but slender and creeping, stems often unbranched. Leaves usually papyraceous when dry, up to 10(19) cm. long; lateral veins 4–10(12) pairs; petiole up to 2(3) cm. long. Peduncle 0.5–1(2) cm. long. Flowering face subcircular, ± quadrangular to substellate; appendages of the inner (marginal) row forming a distinct dentate to crenate rim, the appendages occasionally subulate and up to 2 mm. long.

Mozambique. MS: Beira, *Schlechter* s.n. (B).

Also in Tanzania, Kenya, Uganda, Rwanda, Burundi and Zaire. In coastal forest and woodland. Var. *hildebrandtii* occurs in eastern Tanzania and Kenya in drier habitats than the more mesic var. *schlechteri*. It is recognised by its stems which are usually branched, thickly succulent and often swollen and tuber-like at the base. It may also be distinguished by petioles up to 0.5(1) cm. long, peduncles 0.2–0.8(1.2) cm. long, and by the flowering face of the receptacle which may be subcircular to ± angular or elliptic.

6. **Dorstenia buchananii** Engl., Bot. Jahrb. **20**: 142 (1894); Mon. Afr. Pflanzen. **1**: 23 (1898). —Rendle in F.T.A. **6**, 2: 48 (1916). —Hijman in F.T.E.A., Moraceae: 36 (1989). Type: Malawi, without locality, *Buchanan* 505 (B, holotype; K, isotype).

Herb, tuberous; tuber discoid to subglobose up to 4.5 cm. in diam.; stems annual, erect, ascending or creeping and rooting at the nodes, up to 50(90) cm. long, 1–7(10) mm. thick, puberulous, usually branched, the branches often arrested (especially on creeping stems), the arrested branches transformed into 1 cm. wide, subglobose tubers. Leaves spirally arranged, those on the lower part of the stem usually entire to tripartite, scale leaves; lamina oblong to elliptic or subobovate, sometimes subcircular, (0.5)2–15 × (0.3)1–5 cm., papyraceous when dry, apex obtuse to acute or subacuminate to rounded, base cuneate to attenuate or obtuse, margin ± irregularly coarsely dentate-crenate to denticulate; superior surface smooth, puberulous to strigillose, the inferior surface puberulous; lateral veins 4–12(14) pairs; petiole 3–20 mm. long, 1–2 mm. thick; stipules narrowly triangular to subulate, up to 1.5(4) mm. long, persistent. Inflorescences solitary; peduncle 0.2–1(1.6) cm. or 4–9 cm. long; receptacle broadly turbinate, and elliptic to subnaviculate in outline, 1–2.3 × 0.5–1(1.2) cm.; flowering face narrowly elliptic, margin very narrow or lacking; appendages in 2 rows, the inner (marginal) row of semi-circular appendages, up to 0.5(1) mm. long, forming a crenate rim, the outer (submarginal) row of 2 terminal appendages, band-shaped to filiform, 15–100(130) mm. long. Staminate flowers crowded, perianth lobes 2, stamens 2. Pistillate flowers numerous; perianth tubular, stigmas 2. Endocarp body tetrahedral, c. 1.5 mm. in diam., tuberculate.

Var. **buchananii** TAB. **15**, fig. A1.

    Dorstenia caudata Engl., Bot. Jahrb. **28**: 377 (1900). —Rendle in F.T.A. **6**, 2: 67 (1916). Type from Tanzania.

    Dorstenia unicaudata Engl., Bot. Jahrb. **51**: 432 (1914). Type from Tanzania.

Stems erect or creeping. Leaf lamina elliptic, 2–6 × 1.5–3.5 cm. Peduncle c. 1 cm. long. Receptacle c. 1.5–2 × 1 cm.; upper appendage c. 80 mm. long, lower appendage c. 15 mm. long.

Zambia. E: Mwangazi R. Valley, fl. 26.xi.1958, *Robson* 727 (BM; K; LISC; SRGH). **Zimbabwe**. E: Mutare, E. side of Municipal Sandpits, 8.iii.1960, *Chase* 7310 (B; BR; FHO; K; LISC; SRGH; UPS; W; WAG). **Malawi**. N: Rumphi/Mzimba Distr., Njakwa Gorge, fl. 30.xii.1973, *Pawek* 7660 (K; SRGH). C: Chitala R. bridge on Chitala-Salima road, fl. 13.ii.1959, *Robson* 1584 (BM; K; LISC; SRGH). **Mozambique**. N: Murrupula, Namacorro area, fl. 26.i.1961, *M.F. Carvalho* 461a (K). MS: Manica, Macequece, fl. 3.i.1948, *Mendonça* 3576 (LISC; MO).

Also in south Tanzania. In shade in evergreen gully forest and riverine forest and deciduous woodland, often in rocky places; up to 1700 m.

36

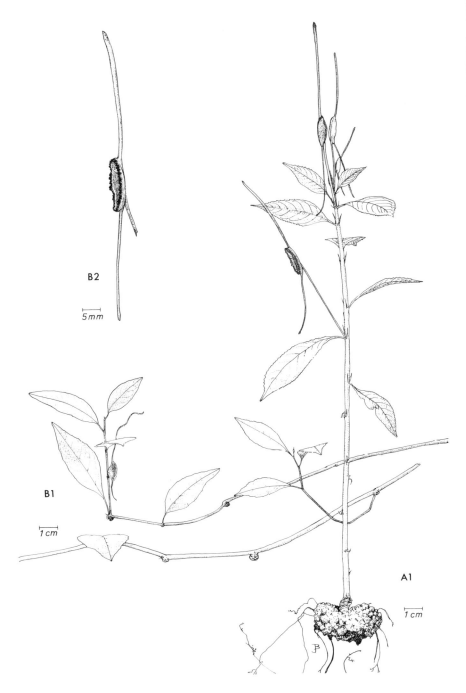

Tab. 15. A.—DORSTENIA BUCHANANII var. BUCHANANII. A1, habit, *Chase* 4311. B. — DORSTENIA BUCHANANII var. LONGIPEDUNCULATA. B1, habit; B2, inflorescence, B1–2 *Milne-Redhead & Taylor* 7739. Drawn by J. Brinkman.

Var. **longipedunculata** Rendle in Journ. Bot. **53**: 302 (1915); in F.T.A. **6**, 2: 48 (1916). —Hijman in F.T.E.A., Moraceae: 38 (1989). TAB. **15**, figs. B1–B2. Type: Mozambique, near Lake Malawi (Nyassa), *Johnson* 494 (K, holotype).
> *Dorstenia ruahensis* Engl., Bot. Jahrb. **28**: 377 (1900). —Rendle in F.T.A. **6**, 2: 67 (1916). Type from Tanzania.
> *Dorstenia longipedunculata* De Wild., Pl. Bequaert. **6**: 44 (1932) as "*longepedunculata*". Type from Tanzania.

Stems erect. Leaf lamina elliptic to oblong (5)6–15 × (1)2–5 cm. Peduncle 4–9 cm. long. Receptacle c. 1–1.3 × 0.5–1 cm.; upper appendage 80–130 mm. long, lower appendage 15–70 mm. long.

Zambia. N: Mbala Distr., Lufubu R., Iyendwe Valley, fl. 7.xii.1959, *Richards* 11888 (K; LISC; SRGH). **Malawi**. S: Mwanza Distr., at Neno, fl., *Salubeni* 330 (SRGH). **Mozambique**. N: Near Lake Malawi (Nyassa), 1902, *Johnson* 494 (K).
Also in Tanzania. In *Brachystegia* woodland, in rocky places; up to 1700 m.

7. **Dorstenia benguellensis** Welw. in Trans. Linn. Soc. **27**: 71 (1869). —Bur. in DC., Prodr. **17**: 274 (1873). —Rendle in F.T.A. **6**, 2: 64 (1916). —Hauman in F.C.B. **1**: 74 (1948). —Hijman in Fl. Cameroun **28**: 94, t. 32, fig. 5–7 (1985); in F.T.E.A., Moraceae: 38 (1989). Type from Angola.
> *Dorstenia poggei* Engl., Bot. Jahrb. **20**: 146 (1894). Type from Zaire.
> *Dorstenia debeerstii* De Wild. & T. Durand in Bull. Soc. Roy. Bot. Belg. **39**: 75 (1900). —Rendle in F.T.A. **6**, 2: 62 (1916). Type from Zaire.
> *Dorstenia verdickii* De Wild. & T. Durand in Bull. Soc. Roy. Bot. Belg. **40**: 26 (1901). Type from Zaire.
> *Dorstenia hockii* De Wild. in Bull. Jard. Bot. Brux. **3**: 278 (1911). Type from Zaire.
> *Dorstenia wellmannii* Engl., Bot. Jahrb. **46**: 276 (1911). Type from Angola.
> *Dorstenia stenophylla* R.E. Fries in Ark. Bot. **13**, 1: 9, t. 2, fig. 4 (1913). Type: Zambia, Kalambo, *R.E. Fries* 1332 (UPS, holotype).
> *Dorstenia rhodesiana* R.E. Fries in Ark. Bot. **13**, 1: 11, t. 2, fig. 1, 2 (1913). Type: Zambia, Msisi, *R.E. Fries* 1302 (UPS, holotype).
> *Dorstenia mirabilis* R.E. Fries in Ark. Bot. **13**, 1: 13, t. 2, fig. 3 (1913). Type: Zambia, Kuta, *R.E. Fries* 1082 (UPS, holotype).
> *Dorstenia rosenii* R.E. Fries in Ark. Bot. **13**, 1: 15, t. 2, fig. 5 (1913). Type: Zambia, Mukanshi R., *R.E. Fries* 1125 (UPS, holotype).
> *Dorstenia rosenii* var. *multibracteata* R.E. Fries in Ark. Bot. **13**, 1: 15 (1913). Type: Zambia, Mbala, *R.E. Fries* 1125a (UPS, holotype).
> *Dorstenia sessilis* R.E. Fries in Ark. Bot. **13**, 1: 16, t. 2, fig. 6–8 (1913). Type: Zambia, Mukanshi, between Lake Bangweulu and Tanzania (UPS, holotype; B, isotype).
> *Dorstenia katangensis* De Wild. in Fedde, Repert. **11**: 519 (1913). Type from Zaire.
> *Dorstenia piscicelliana* Buscal. & Muschler in Engl., Bot. Jahrb. **49**: 464 (1913). Type: Zambia, between Kabwe and Bwana Mukubwa (Buana Mukuba) (B, holotype).
> *Dorstenia homblei* De Wild. in Fedde, Repert. **13**: 195 (1914). Type from Zaire.
> *Dorstenia poggei* var. *meyeri-johannis* Engl., Bot. Jahrb. **51**: 434 (1914). Type from Tanzania.
> *Dorstenia debeerstii* var. *multibracteata* (R.E. Fries) Rendle in F.T.A. **6**, 2: 63 (1916). Type as for *Dorstenia rosenii* var. *multibracteata*.
> *Dorstenia mildbraediana* Peter in Fedde, Repert., Beih. **40**, 2: 79, Descr.: 7, t. 8, fig. 1 (1932). Lectotype from Tanzania, chosen by Hijman in F.T.E.A., Moraceae: 40 (1989).
> *Dorstenia achtenii* De Wild., Contr. Fl. Katanga, Suppl. 4: 100 (1932) as "*achteni*". —Hauman in F.C.B. **1**: 75 (1948). Type from Zaire.
> *Dorstenia katubensis* De Wild., Contr. Fl. Katanga, Suppl. 4: 102 (1932). Type from Zaire.
> *Dorstenia lactifera* De Wild., Contr. Fl. Katanga, Suppl. 4: 105 (1932). Type from Zaire.
> *Dorstenia stipitata* De Wild., Contr. Fl. Katanga, Suppl. 4: 108 (1932). Type from Zaire.
> *Dorstenia papillosa* Hauman in Bull. Jard. Bot. Brux. **19**: 59 (1948). Type from Zaire.
> *Dorstenia papillosa* var. *plethorica* Hauman in Bull. Jard. Bot. Brux. **19**: 60 (1948). Type from Zaire.
> *Dorstenia campanulata* Hauman in Bull. Jard. Bot. Brux. **19**: 61 (1948). Type from Zaire.
> *Dorstenia achtenii* var. *laxiflora* Hauman in F.C.B. **1**: 76 (1948). Type from Zaire.
> *Dorstenia benguellensis* var. *capillaris* Hauman in F.C.B. **1**: 75 (1948). Type from Zaire.
> *Dorstenia verdickii* var. *scaberrima* Hauman in F.C.B. **1**: 76 (1948). Type from Zaire.

Herb up to 50(60) cm. tall, tuberous; tuber discoid to subglobose, 1–12 cm. in diam. Stems annual, unbranched or branched (branches usually arrested), 1.5–8 mm. thick, puberulous to hirtellous or hispidulous. Leaves spirally arranged, leaves on the lower part of the stem scale-like and entire to tripartite; lamina of normal leaves oblong to subovate or linear, sometimes subobovate, occasionally to elliptic or ovate, 1–15 × 0.2–4.5 cm., chartaceous when dry; apex acute to subacuminate; base obtuse to rounded; margin finely to coarsely dentate or subcrenate; both surfaces puberulous to hirtellous or hispidulous; lateral veins 4–12 (in linear leaves up to 25) pairs; petiole (0)1–2(5) mm. long,

c. 1 mm. thick; stipules triangular to oblong, up to 5 mm. long, persistent. Inflorescences solitary or in pairs; peduncle 0.5–7 cm. long, 1–1.5 mm. thick. Receptacle discoid to broadly turbinate to shallowly cup-shaped, subcircular to elliptic, 0.5–2(2.5) cm. in diam.; margin very narrow; appendages in 2 rows, inner (marginal) row with numerous triangular to subulate, or ovate appendages up to 1(2) mm. long, forming a subdentate rim, or appendages triangular to ovate, up to 7 mm. long forming a subcrenate rim, or appendages indistinct and the rim subentire; outer (submarginal) row usually with c. 10–20 appendages, less commonly to 2 or 1(0), outer appendages ligulate to filiform or subspathulate to oblong, (1)2–20(80) mm. long. Staminate flowers crowded; perianth lobes 2; stamens 2. Pistillate flowers numerous, perianth shortly tubular; stigmas (1)2. Endocarp body tetrahedral, c. 2 mm. in diam., tuberculate (pale brown).

Zambia. B: Kabompo R., 23.x.1966, *Leach & Williamson* 13452 (K; PRE; SRGH). N: Kasama Distr., 28 km. SE. of Kasama, 22.xi.1960, *Robinson* 4101 (EA; K; SRGH). W: Mwinilunga Distr., c. 0.8 km. S. of Matonchi Farm, fl. 24.x.1937, *Milne-Redhead* 2936 (BR; K; LISC). C: Lusaka, Mt. Makulu, fl. 11.i.1972, *Rees* 1068 (SRGH). E: Lundazi Distr., W. of Chikomeni, fl. 3.xii.1970, *Sayer* 495 (SRGH). **Zimbabwe.** N: c. 24 km. W. of Guruve (Sipolilo), fl. 9.xii.1973, *Cannell* 577 (K; SRGH). **Malawi.** N: Mzimba Distr., Mzuzu, fl. 6.xii.1972, *Pawek* 6069 (K; SRGH). C: Kasungu Distr., Chipala Hill, Chamama, fl. 16.i.1959, *Robson* 1216 (BM; K; LISC; SRGH). **Mozambique.** Z: Namacurra to Milange, fl. 3.ii.1966, *Torre & Correia* 14402 (LISC).

Also in Angola, Tanzania, Kenya, Rwanda, Burundi, Uganda, Sudan, Zaire, Central African Republic and Cameroon. In miombo and mixed deciduous woodland, often on rocky slopes, in shallow or sandy soils, Kalahari Sand, and seasonally waterlogged grasslands (dambos); up to 2450 m.

8. **Dorstenia cuspidata** A. Rich., Tent. Fl. Abyss. **2**: 272 (1851). —Hochst. in Flora **27**: 103 (1844). —Bur. in DC., Prodr. **17**: 275 (1873). —Engl., Mon. Afr. Pflanzen. **1**: 21 (1898). —Rendle in F.T.A. **6**, 2: 68 (1916). —Perrier & Léandri in Humbert, Fl. Madag., Moraceae: 20 (1953). —Hijman in Fl. Cameroun **28**: 92, t. 32 figs. 9, 10 (1985); in F.T.E.A., Moraceae: 40 (1989). Type from Ethiopia.

Herb 10–40(50) cm. tall, tuberous or with swollen rhizomes; tuber discoid to subglobose or irregular in shape, 1–4 cm. in diam. Stems annual, mostly unbranched, 1.4 mm. thick, puberulous. Leaves spirally arranged, leaves on lower part of stem scale-like and entire to tripartite; lamina of normal leaves elliptic to oblong or subobovate to lanceolate, sometimes linear, 1.5–12 × 0.5–5.5 cm.; papyraceous to chartaceous when dry, apex obtuse to acute; base cuneate to attenuate; margin finely crenate to dentate, coarsely dentate-crenate or repand to subentire; superior surface puberulous, substrigillose to subhispidulous; inferior surface puberulous to hirtellous; lateral veins 3–12 (or in linear leaves up to 25) pairs; petiole (2)5–25 mm. long, 1–1.5 mm. thick; stipules narrowly triangular to subulate, (0.5)2–5 mm. long, persistent. Inflorescences solitary or in pairs; peduncle 0.8–3(7) cm. long, 0.5–1 mm. thick; receptacle discoid to broadly turbinate, irregularly stellate to subquadrangular, triangular to subcircular or sometimes subnaviculate, 0.7–2.2 cm. in diam., or if subnaviculate, 1–1.5 × 0.3–0.5 cm.; flowering face ± similar to the receptacle in outline, margin very narrow to absent; appendages in two rows. Appendages of the inner (marginal) row numerous, triangular to ovate, up to 1 mm. long, forming a crenate rim, or indistinct and the rim subentire; appendages of the outer (submarginal) row 2 in subnaviculate receptacles or 3, in triangular receptacles, or usually 4–14 in narrowly triangular to band-shaped or filiform receptacles, (5)20–60(85) mm. long. Staminate flowers crowded to ± spaced; perianth lobes 2; stamens 2. Pistillate flowers numerous; perianth shortly tubular; stigma 1. Endocarp body tetrahedral, 1.2–1.5 mm. in diam., tuberculate, greyish.

Var. **cuspidata** —Hijman in Kew Bull. **45**: 362 (1990).
    Dorstenia walleri Hemsley in Gard. Chron., ser. 3, **14**: 178 (1893). —Rendle in F.T.A. **6**, 2: 68 (1916). —Peter in F.W.T.A., ed. 2, **1**: 599 (1958). Type: a plant cultivated at Kew from seeds collected by *Buchanan* in Malawi, Shire Highlands, Nakatupa, (K, holotype).
    Dorstenia caulescens Schweinf. ex Engl., Bot. Jahrb. **20**: 144 (1894) non L. (1753). Type from Sudan.
    Dorstenia unyikae Engl. & Warb., Bot. Jahrb. **30**: 291 (1901). Type from Tanzania.
    Dorstenia gourmaensis A. Chev. in Bull. Soc. Bot. Fr. **58**, Mém. 8d: 207 (1912). Type from Burkina Fasso.
    Dorstenia gourmaensis var. *floribunda* A. Chev. in Bull. Soc. Bot. Fr. **58**, Mém. 8d: 207 (1912). Type from Burkina Fasso.
    Dorstenia walleri var. *minor* Rendle in F.T.A. **6**, 2: 68 (1916). Type as for *Dorstenia unyikae*.

Dorstenia tetractis Peter in Fedde, Repert., Beih. **40**, 2: 80, Descr.: 8, t. 10, fig. 1 (1932). Type from Tanzania.

Dorstenia quarrei De Wild. & Staner, Contr. Fl. Katanga, Suppl. 4: 106 (1932). Type from Zaire.

Succulent herbs to 40 cm. tall, stems erect from globose or discoid tubers 1–4 cm. across. Leaves with 8–12 (25 in linear leaves) pairs of lateral veins. Receptacle broadly turbinate, stellately lobed with (3)4–8(14) lobes, and (3)4–8(14) appendages up to 40 mm. long.

Zambia. B: Mongu, Lealui, Likone For. Res., Mongu to Kaoma, mile 30, 26.xii.1952, *White* 2140 (FHO; K). N: Chinsali Distr., Machipara Hill, Shiwa-Ngandu, 16.i.1959, *Richards* 10681 (BR; EA; K; LISC). W: Copperbelt, Mpongwe-Lake Kashiba road, 23.ii.1982, *Berg & Bingham* 1396 (K; U). C: Lusaka-Kabwe road, 19.ii.1982, *Berg & Bingham* 1380 (K; U). E: Sasare, Copper Mine, 9.xii.1958, *Robson* 874 (K). S: Kalomo Distr., Kasakwi Forest, Simamba, Sichifulo Controlled Area, 11.ii.1965, *Mitchell* 26/10 (K; SRGH). **Zimbabwe**. N: Guruve Distr., Semhenheka, c. 11 km. S. of Guruve, 9.xii.1970, *Mavi* 1197 (K; LISC; SRGH). **Malawi**. N: Mzimba Distr., Lunyangwa R., c. 1.6 km. SW. on cut off to M1 from Mzuzu, 18.i.1978, *Pawek* 13595 (K). C: Dedza Distr., Mua-Livulezi For. Res., Kamwanje Stream, 5.i.1965, *Banda* 598 (K; SRGH). S: River Shire, near Liwonde Ferry, 13.iii.1955, *Exell, Mendonça & Wild* 846 (LISC; SRGH). **Mozambique**. N: Ribáuè, 27.i.1964, *Torre & Paiva* 10259 (LISC). Z: Quelimane Distr., Moebede road, 26.xi.1948, *Faulkner* 330 (BR; K).

Also in Tanzania and Ethiopia, and from Sudan to Senegal. Deciduous or miombo woodland and riverine vegetation, on rocky outcrops and dambo margins, often on termitaria; up to 1500 m.

Var. **brinkmaniana** Hijman in F.T.E.A., Moraceae: 40 (1989); in Kew Bull. **45**: 364 (1990). Type from Tanzania.

Succulent plants up to 35 cm. tall, with erect stems arising from globose (discoid) tubers 1.5–3 cm. across. Leaves with 5–8 pairs of lateral veins. Receptacle discoid, subtriangular to angular or subcircular with 3–10 appendages, (5)20–60(85) mm. long.

Zambia. N: Mbala Distr., Chimka Bay, Lake Tanganyika, 28.xii.1963, *Richards* 18705 (BR; K; UPS). Also in Tanzania and Kenya. Thicket vegetation (dry forests or woodland in E. Africa), often in rocky places.

Var. *humboltiana* (Baill.) Léandri is endemic to Madagascar, while var. *preussii* (Engl.) Hijman occurs in Senegal, Sierra Leone, Mali, Guinée, Guinea Bissau, Nigeria, Cameroon, Gabon and Ethiopia

Note: **Dorstenia barnimiana** Schweinf. would be expected to occur in the Mbala District of Zambia. It is recorded from the Ufipa District of Tanzania and northwards to the Cameroons, Ethiopia and the Arabian Peninsula. It may be distinguished by its petioles and peduncles arising directly from the apex of a tuber; its ± rosulate leaves and short internodes; and its vertical, naviculate or subligulate receptacle.

10. FICUS L.

Ficus L., Sp. Pl.: 1059 (1753); Gen. Pl., ed. 5: 482 (1754). —Corner in Gard. Bull., Singapore **21**: 1 (1965). —C.C. Berg in F.T.E.A., Moraceae: 43 (1989); in Kirkia **13**: 253–291 (1990). —van Greuning in S. Afr. J. Bot. **56**, 6: 599–630 (1990).

Urostigma Gasp., Nova Gen. Fici: 7 (1844).
Galoglychia Gasp., Nova Gen. Fici: 10 (1844).
Sycomorus Gasp., Ricerche Caprifico: 86 (1845).
Pharmacosycea Miq. in Hook., Lond. Journ. Bot. **6**: 525 (1847).

Trees, shrubs or sometimes lianas, terrestrial or hemi-epiphytic (and then with aerial roots, often strangling and secondarily terrestrial), monoecious or (gyno)dioecious. Leaves spirally arranged, distichous, subopposite or subverticillate; lamina pinnately veined, with glandular (sometimes waxy) spots in the axils of at least the basal lateral veins beneath or (mostly a single spot) at the base of the midrib beneath; stipules fully amplexicaul, semi-amplexicaul or lateral, free or partly connate. Inflorescences (figs or syconia) solitary or in pairs in the leaf axils, mostly several together on small spurs in the leaf axils down to the lesser branches or on distinct spurs on the lesser branches down to the trunk, or on leafless branchlets on the older wood, down to the base of the trunk. Figs consist of an urceolate receptacle with a ± narrow apical opening (ostiole), the flowers enclosed within; figs are functionally bisexual, containing staminate flowers, long styled flowers (destined to produce seeds) and short styled flowers (destined to hatch the larvae of the pollinator); or the figs on one plant occur as "gall figs" containing staminate

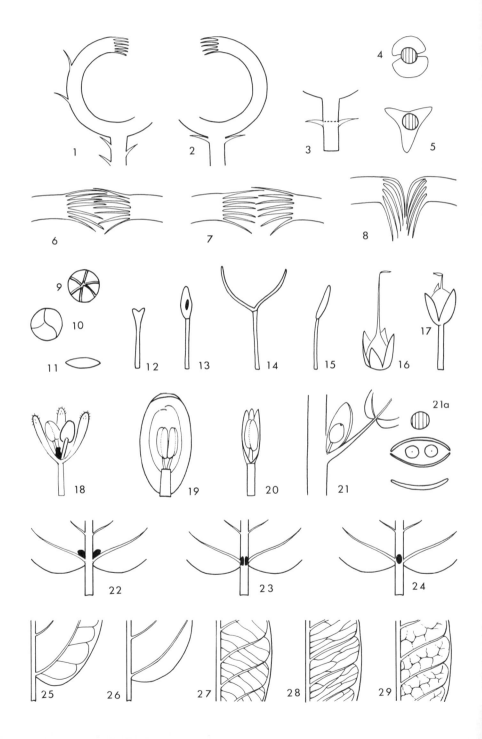

flowers and short styled flowers (which do not produce seeds but serve as breeding sites for the pollinator) and on another plant occur as "seed figs" containing only long-styled, fully functional, pistillate flowers. Bracts 2–3(4) on the peduncle (peduncular bracts) or 2–3 subtending the receptacle (basal bracts), sometimes on the outer surface of the receptacle (lateral bracts); numerous bracts in the opening of the receptacle (ostiolar bracts), and often among the flowers (interfloral bracts). Staminate flowers occur near the ostiole or dispersed within the inflorescence, mostly pedicellate; tepals 2–6, free or connate; stamens 1–3, pistillode mostly absent. Pistillate flowers sessile or pedicellate; tepals 2–6(7), free or connate; ovary free, stigmas 1 filiform or infundibuliform or 2 and ± filiform or subulate, the stigmas cohering; pistillate flowers in (gyno)dioecious taxa are strongly differentiated into long-styled and short-styled flowers, while in monoecious taxa (where all pistillate flowers set seed) styles of intermediate lengths often occur. Fruits achene-like or more often ± drupaceous, releasing the endocarp body (pyrene) or not; at the fruiting stage the wall of the fig becomes ± fleshy, yellow, orange, red, pink or purplish, or green. "Gall fruits" achene-like.

The genus comprises c. 750 species with c. 500 species in Asia and Australasia, c. 150 species in the Neotropics and c. 100 species in Africa and vicinity.

The present subdivision of the genus was proposed by Corner in Gard. Bull., Singapore **21**: 3–6 (1965) with full synonymy. The principal diagnostic features are shown in TAB. **16**.

Ficus species not yet found in the Flora Zambesiaca area, but recorded from localities near its borders are *F. amadiensis* De Wild. from Kyimbila in Tanzania, *F. conraui* Warb. and *F. tesselata* Warb. from Shaba in Zaire and *F. bizanae* Hutch. from Natal, in S Africa.

Key to Cultivated Species

The following species have been introduced from Asia and Western Africa.

1. Root climber; leaves heteromorphic on sterile and fertile branches; figs pear-shaped,
   up to 5(7) cm. long   -   -   -   -   -   -   -   -   -   - *Ficus pumila* L.
 – Trees or shrubs; leaves not heteromorphic   -   -   -   -   -   -   -   - 2
2. Leaf lamina 3–5-lobed; figs ± broadly pear-shaped, pedunculate   -   - *Ficus carica* L.
 – Leaf lamina entire; figs globose to ellipsoid, sessile   -   -   -   -   -   - 3
3. Leaf lamina pandurate, usually longer than 30 cm.   -   -   -   - *Ficus lyrata* Warb.
 – Leaf lamina not pandurate, usually shorter than 20 cm.   -   -   -   -   -   - 4

Tab. 16. Diagrams of key features in FICUS. 1, receptacle with peduncular, lateral and ostiolar bracts (sect. *Sycidium*); 2, receptacle with basal and ostiolar bracts (other sections); 3, stipitate receptacle; 4, basal bracts two (sect. *Galoglychia*); 5, basal bracts three (subgen. *Sycomorus*, sect. *Oreosycea*, sect. *Urostigma*); 6, most ostiolar bracts interlocking (sect. *Sycidium*, subgen. *Sycomorus*, sect. *Urostigma*); 7, the middle ostiolar bracts not interlocking (sect. *Oreosycea*); 8, all ostiolar bracts descending (sect. *Galoglychia*); 9, ostiole circular, several ostiolar bracts visible (sect. *Sycidium*, subgen. *Sycomorus*); 10, ostiole circular, three ostiolar bracts visible (sect. *Oreosycea*, sect. *Urostigma*); 11, ostiole slit-shaped (sect. *Galoglychia*); 12, stigma ± infundibuliform (sect. *Sycidium*, subgen. *Sycomorus*); 13, stigma ± flame-shaped and slightly infundibuliform (subgen. *Sycomorus*); 14, stigmas two (sect. *Oreosycea*, and sometimes in sect. *Galoglychia*); 15, stigma one, elongate (sect. *Urostigma*, sect. *Galoglychia*); 16, long-styled ("seed") flower, often sessile or short-pedicellate; 17, short-styled ("gall") flower, ± long-pedicellate; 18, staminate flower with distinct pistillode and hairy tepals (sect. *Sycidium*); 19, staminate flower with saccate perianth and two stamens, being enveloped by two large bracts or bracteoles (subgen. *Sycomorus*); 20, staminate flower with one stamen (sect. *Oreosycea*, sect. *Urostigma*, sect. *Galoglychia*); 21, calyptrate bud-cover (as found in *F. craterostoma* and *F. ovata*); 21a, diagram of same; 22, glandular spots in the axils of the basal (or main) lateral veins beneath (sect. *Sycidium*, subgen. *Sycomorus*); 23, glandular spots two, on the base of the midrib beneath (sect. *Oreosycea*); 24, glandular spot one, on the base of the midrib beneath (sect. *Urostigma*, sect. *Galoglychia*); 25, basal lateral veins branched; 26, basal lateral veins unbranched; 27, tertiary venation scalariform; 28, tertiary venation predominantly parallel to the lateral veins; 29, tertiary venation reticulate. Redrawn by S. Dawson.

4. Leaf lamina with 15 or more pairs of lateral veins; stipules long; figs
   ellipsoid   -   -   -   -   -   -   -   -   -   -   -   - *Ficus elastica* Roxb.
 – Leaf lamina with at most 10 pairs of lateral veins; stipules short; figs mostly
   globose   -   -   -   -   -   -   -   -   -   -   -   -   -   -   - 5
5. Leaf base truncate, apex caudate, margin repand; petiole long and
   slender   -   -   -   -   -   -   -   -   -   -   -   -   - *Ficus religiosa* L.
 – Leaf base rounded to cuneate, apex ± rounded or acute, margins entire; petiole short or
   stout   -   -   -   -   -   -   -   -   -   -   -   -   -   -   - 6
6. Leaf lamina usually longer than 10 cm.; petiole usually longer than 2 cm.; figs 1–1.5 cm.
   in diam.   -   -   -   -   -   -   -   -   -   -   - *Ficus benghalensis* L.
 – Leaf lamina less than 10 cm. long; petiole up to 2 cm. long; figs 0.5–1 cm.
   in diam.   -   -   -   -   -   -   -   -   -   -   -   -   - 7
7. Leaf apex distinctly acuminate; figs turning orange to yellow
   or red   -   -   -   -   -   -   -   -   -   -   -   - *Ficus benjamina* L.
 – Leaf apex shortly and faintly acuminate; figs turning purple or
   blackish   -   -   -   -   -   -   -   -   -   - *Ficus microcarpa* L.f.

Key to sections of Ficus subgenera in the Flora Zambesiaca area

1. Ostiole slit-shaped (TAB. 16, fig. 11), sometimes tri-radiate; all ostiolar bracts descending
   (TAB. 16, fig. 8)   -   -   -   -   -   -   -   -   -   - Sect. **Galoglychia,** *p. 44*
 – Ostiole circular; at least 3 ostiolar bracts visible, none (or only the lower bracts) descending
   (TAB. 16, figs. 1, 2, 6, 7, 9, 10)   -   -   -   -   -   -   -   -   -   - 2
2. More than 3 ostiolar bracts visible (TAB. 16, fig. 9); glandular spots present in the axils of the
   main basal lateral nerves on leaf lower surface (TAB. 16, fig. 22)   -   -   -   - 3
 – Three ostiolar bracts visible (TAB. 16, fig. 10); 1–2 glandular spots borne at the base of the
   midrib, on the leaf lower surface (TAB. 16, figs. 23, 24)   -   -   -   -   -   - 4
3. Stipules semi-amplexicaul to lateral, not enclosing the stem apex; dioecious trees
   or shrubs   -   -   -   -   -   -   -   -   -   - Sect. **Sycidium,** *p. 42*
 – Stipules fully amplexicaul, completely enclosing the stem apex; monoecious
   trees   -   -   -   -   -   -   -   -   -   -   - Sect. **Sycomorus,** *p. 43*
4. Lenticels concentrated on the uppermost part of the internodes; stigmas 2
   (TAB. 16, fig. 14)   -   -   -   -   -   -   -   -   -   - Sect. **Oreosycea,** *p. 43*
 – Lenticels scattered over the internodes; stigmas 1(2) (TAB. 16, fig. 15)
   -   -   -   -   -   -   -   -   -   -   -   -   -   - Sect. **Urostigma,** *p. 43*

Subgen. FICUS

Sect. **SYCIDIUM** Miq.

Shrubs or trees, terrestrial, (gyno)dioecious; sap watery. Leaves distichous and alternate, or subopposite or subverticillate, often lobed to divided, especially when juvenile, margin dentate or subentire; tertiary venation scalariform to reticulate; glandular spots in the axils of the main basal lateral veins on leaf lower surface; stipules lateral to semi-amplexicaul, free. Figs often solitary in the leaf axils or on minute spurs on the older wood, pedunculate, with 2–4 bracts on the peduncle plus several lateral bracts; ostiole circular with several ostiolar bracts visible, only the lower ostiolar bracts descending; interfloral bracts few or absent. Staminate flowers located near the ostiole; tepals 3–6, at least ciliolate; stamens 1–3; pistillode usually present. Pistillate flowers distinctly different with regard to the style length; tepals 4–6, at least ciliolate; stigma 1; endocarp body often released. Wall of the fruiting fig soft, red, orange, or yellow.

This section comprises some 105 species, the majority of which occur in Asia. Four species are found in continental Africa, and four in Madagascar the Mascarene Islands and the Seychelles.

Key to the species of Sect. Sycidium

1. Leaf asymmetrical, lateral veins 3–10-paired; peduncle up to 4 mm.
   long   -   -   -   -   -   -   -   -   -   -   -   -   -   - 2. *asperifolia*
 – Leaf symmetrical, or if asymmetrical then the lateral veins 3–5-paired; peduncle at least 5 mm.
   long   -   -   -   -   -   -   -   -   -   -   -   -   -   -   - 2
2. Leaf lateral veins 3–5(6)-paired; trees   -   -   -   -   -   -   - 1. *exasperata*
 – Leaf lateral veins 5–12-paired; shrubs   -   -   -   -   -   -   -   -   - 3
3. Leaves mostly subopposite or subverticillate; stipules 5–10 mm. long, subpersistent; leaf margin
   usually subentire to faintly crenate, often 3-dentate at the apex   -   -   - 4. *capreifolia*
 – Leaves alternate; stipules 2–6 mm. long, caducous; leaf margin usually serrate-dentate or
   pinnately incised   -   -   -   -   -   -   -   -   -   - 3. *pygmaea*

Subgen. SYCOMORUS (Gasp.) Mildbr. & Burret

Sect. **SYCOMORUS** (Gasp.) Miq.

Trees or less often shrubs, terrestrial, monoecious; sap milky. Leaves spirally arranged or tending to be distichous; margin mostly dentate, crenate or repand; tertiary venation mostly scalariform; glandular spots in the axils of the main basal lateral veins on the leaf lower surface; stipules fully amplexicaul, free. Figs solitary (or in pairs) in the leaf axils, or 1–3 together on unbranched or branched leafless branchlets on the older wood, often down to the trunk, pedunculate, basal bracts 3; lateral bracts absent; receptacle rather large; ostiole circular with several ostiolar bracts visible, only the lower ones descending; interfloral bracts absent from among the pistillate flowers. Staminate flowers located near the ostiole, at first enveloped by 2(3) bracteoles; perianth saccate, 3-lobed; stamens 2–3. Pistillate flowers long-styled or short-styled but intermediate style lengths also occur; tepals 2–4–6, irregularly shaped, free or connate; stigma 1; endocarp body often released. Wall of the fig at fruit stage soft, red to orange or yellowish.

This section comprises 13 species, of which 5 occur in continental Africa, with one extending into Madagascar. Seven species occur in Madagascar and/or the Comoro and Mascarene Islands. One species occurs in Asia.

Key to the species of Sect. Sycomorus

1. Hairs of the petioles and leafy twigs conspicuously of 2 kinds (very numerous short hairs interspersed with scattered, much longer white to yellowish hairs especially on the nodes); leaves usually scabrous above, sometimes hispidulous to glabrescent - - - - - - - - - - - - - - - 5. *sycomorus*
 – Hairs of the petioles and leafy twigs not conspicuously different in length, or absent; leaves usually smooth above, sometimes scabrous - - - - - - - - - 2
2. Peduncles 4–6 mm. in diam.; receptacle 3–6(10) cm. in diam. when fresh, 1–5 cm. in diam. when dry; periderm of the leafy twig (when dry) flaking off; leaf lamina usually as long as wide - - - - - - - - - - - - - - 7. *vallis choudae*
 – Peduncles 1–3 mm. in diam.; receptacle 2–4 cm. in diam. when fresh, 0.5–2.5 cm. in diam. when dry; periderm of the leafy twig (when dry) usually not flaking off; leaf lamina usually distinctly longer than wide - - - - - - - - - - - - - 6. *sur*

Subgen. PHARMACOSYCEA (Miq.) Miq.

Sect. **OREOSYCEA** (Miq.) Miq.

Trees or less often shrubs, terrestrial, monoecious; sap watery, not abundant. Lenticels concentrated on the uppermost part of the internodes. Leaves spirally to distichously arranged, margin entire to dentate, when juvenile irregularly pinnately incised, tertiary venation mostly scalariform; glandular spots 1–2 at the base of the midrib on leaf lower surface; stipules fully amplexicaul, when juvenile semi-amplexicaul, free. Figs solitary or in pairs in the leaf axils, pedunculate; basal bracts 3; lateral bracts sometimes present; ostiole circular with 3 ostiolar bracts visible, only the lower ostiolar bracts descending; interfloral bracts lacking in the 2 African species. Staminate flowers located near the ostiole; tepals 2–3, connate, glabrous; stamen 1; pistillode sometimes present. Pistillate flowers long-styled or short-styled but intermediate style lengths also occur; tepals 2–3, connate, glabrous; stigmas 2, ± filiform; fruits achene-like. Wall of the fruiting receptacle firm, yellow to orange.

This section comprises some 55 species, of which two occur in continental Africa and are closely related, and two occur in Madagascar.

Only one species is recorded from the Flora Zambesiaca area, namely: 8. *Ficus dicranostyla*.

Subgen. UROSTIGMA (Gasp.) Miq.

Sect. **UROSTIGMA**

Trees or shrubs, usually terrestrial (or epilithic), monoecious; sap milky. Leaves spirally arranged, margin entire to subentire; tertiary venation reticulate; glandular spot at the base of the midrib on leaf lower surface; stipules fully amplexicaul, free. Figs in pairs or sometimes more together in the leaf axils or just below the leaves, pedunculate or sessile; basal bracts 3; lateral bracts absent; ostiole circular, with 3 ostiolar bracts visible, only the lower ostiolar bracts descending; interfloral bracts absent or few. Staminate flowers usually only near the ostiole; tepals 3–4, free; stamen 1; pistillode sometimes present. Pistillate flowers long-styled or short-styled but intermediate style lengths also occur; tepals 3–4(7), free; stigma 1, ± filiform; fruit achene-like. Wall of the fruiting fig soft, red, pink or purplish.

This section comprises 20 species, of which 15 occur in Asia, and 3 in continental Africa. One species occurs in Madagascar, and one in the Seychelles and Mascarene Islands.

Key to the species of Sect. Urostigma

1. Main basal pair of veins running well above the leaf margin, often with 2–4 strong branches from the lower side (TAB. 16, fig. 25); the main lateral veins forking far from the leaf margin   -   -   -   -   -   -   -   -   -   -   -   -   -   -   -   - 9. *ingens*
– Main basal pair of veins ± closely following the leaf margin, seldom strongly branched; the main lateral veins forking near the leaf margin   -   -   -   -   -   -   -   -   - 2
2. Figs sessile   -   -   -   -   -   -   -   -   -   - 10. *cordata* subsp. *cordata*
– Figs pedunculate   -   -   -   -   -   -   -   -   -   -   -   -   -   - 3.
3. Leaf lamina (3)5(8) times as long as the petiole; stipules up to 1.5 cm. long; figs mostly smooth when dry; tree of dry places   -   -   -   -   -   -   - 10. *cordata* subsp. *salicifolia*
– Leaf lamina (5)8–10 times as long as the petiole; stipules up to 4 cm. long; figs wrinkled when dry; shrub or small tree of wet places   -   -   -   -   -   -   -   -   - 11. *verruculosa*

Sect. **GALOGLYCHIA** (Gasp.) Endl. —C.C. Berg in Proc. Konink. Ned. Akad. Weten., Ser. C, **89**: 121–127 (1986).
Subgen. *Bibracteatae* Mildbr. & Burret in Engl., Bot. Jahrb. **46**: 175 (1911).

Trees, shrubs or sometimes lianas, hemi-epiphytic or terrestrial (or epilithic), mostly with aerial roots; sap milky. Leaves spirally arranged, sometimes almost distichous and/or subopposite, margin entire or almost so, tertiary venation scalariform to reticulate or ± parallel to the lateral veins; glandular spot present at the base of the midrib on leaf lower surface; stipules fully amplexicaul, free or partly connate. Figs in the leaf axils or just below the leaves, or on spurs on the lesser branches, or down to the base of the trunk, pedunculate or sessile; basal bracts 2–3; lateral bracts absent; ostiole slit-shaped or triradiate; all ostiolar bracts descending; interfloral bracts present. Staminate flowers dispersed; tepals 2–4, free or connate, glabrous; stamen 1; pistillode usually absent. Seed and gall flowers ± different; tepals 2–4, free or partly connate, glabrous; stigmas 1(2); fruit achene-like, or drupaceous and then often releasing the endocarp body, or the upper part forming a mucilaginous cap. Wall of the fruiting fig soft to rather firm, reddish, orange, yellow, greenish, purplish, or brownish.

This section is confined to the African floristic region. 72 species are recognised, of which 66 occur in continental Africa with 3 also known from the Indian Ocean Islands.

Six more or less distinct groups (subsections) can be recognised within this section, 5 of which occur in the Flora Zambesiaca area.
Species 12–20 = Group 1, species 21 = Group 2, species 22–27 = Group 3, species 28–34 = Group 4, species 35–38 = Group 6. Group 5 is not represented in the Flora Zambesiaca area.

Key to the species of Sect. Galoglychia

1. Figs with a peduncle at least 2 mm. long   -   -   -   -   -   -   -   -   -   - 2
– Figs sessile or subsessile   -   -   -   -   -   -   -   -   -   -   -   - 35
2. Figs on distinct spurs 5–50(150) mm. long, borne on older wood (in *F. ottoniifolia* sometimes on branchlets up to the leaf axils); figs at least 10 mm. in diam. when dry   -   -   - 3
– Figs not on spurs, or on minute spurs 1–2 mm. long borne in the leaf axils and/or just below the leaves and then figs at most 10 mm. in diam. when dry   -   -   -   -   -   - 15
3. Basal bracts persistent (or subpersistent) (see TAB. 16, figs. 2, 4, 5)   -   -   -   - 4
– Basal bracts caducous   -   -   -   -   -   -   -   -   -   -   -   -   - 11

4. Stipules with short, white, appressed hairs; leaves usually minutely puberulous on the main veins beneath (long stipules on new-flush growth with patent hairs)   -   -   - 31. *chirindensis*
  - Stipules glabrous (or with minute, patent hairs or distinctly ciliolate); leaves (except sometimes midrib) glabrous beneath   -   -   -   -   -   -   -   -   -   -   -   - 5
5. Figs ellipsoid, if globose then at most 15 mm. in diam. when dry, mostly smooth   -   -   -   -   -   -   -   -   -   -   -   -   -   -   - 6
  - Figs globose, 15–30 mm. in diam. when dry, mostly wrinkled   -   -   -   -   - 7
6. Leaf 8–16 cm. long, base acute to obtuse   -   -   - 28. *ottoniifolia* subsp. *ulugurensis*
  - Leaf 2–9 cm. long, base cordulate to rarely rounded   -   -   -   -   - 29. *tremula*
7. Peduncle 5–10(11) mm. long; petiole 2–4 mm. thick when dry; apex of the thickly coriaceous lamina rounded or very shortly and bluntly acuminate   -   -   -   -   - 33. *bubu*
  - Peduncle (9)10–20 mm. long; petiole 1–2(3) mm. thick when dry; apex of the coriaceous to subcoriaceous lamina usually acuminate   -   -   -   -   -   -   -   -   - 8
8. Fig-bearing spurs more than 1 cm. long, mostly 5–15 cm. long, their bud scales ± densely puberulous; midrib and petiole usually drying red-brown   -   -   -   -   -   - 9
  - Fig-bearing spurs up to 1 cm. long, their bud scales glabrous; midrib and petiole usually drying blackish   -   -   -   -   -   -   -   -   -   -   -   -   - 10
9. Stipules ciliolate; fig-bearing spurs usually up to c. 5 cm. long   -   -   -   -   -   -   -   -   -   - 32. *sansibarica* subsp. *sansibarica*
  - Stipules not ciliolate; fig-bearing spurs usually up to c. 10(15) cm. long   -   -   -   -   -   -   -   -   -   - 32. *sansibarica* subsp. *macrosperma*
10. Leaf lateral veins 5–8(9) pairs; figs 2–4 cm. in diam. when dry   -   -   -   -   -   -   -   -   -   -   - 30. *polita* subsp. *polita*
  - Leaf lateral veins (8)10–12 pairs; figs 1–1.5 cm. in diam. when dry   -   -   -   -   -   -   -   -   -   - 30. *polita* subsp. *brevipedunculata*
11. Stipules with short, white, appressed hairs (but not ciliolate); leaf usually minutely puberulous on the main veins beneath (long stipules on new-flush growth with patent hairs)   -   -   -   -   -   -   -   -   -   -   - 31. *chirindensis*
  - Stipules glabrous (or with minute, patent hairs or distinctly ciliolate); leaf (except sometimes midrib) glabrous beneath   -   -   -   -   -   -   -   -   -   -   - 12
12. Figs globose, 15–35 mm. in diam. when dry, mostly wrinkled   -   -   -   -   - 13
  - Figs ellipsoid, if globose then at most 15 mm. in diam. when dry, mostly smooth   -   -   -   -   -   -   -   -   -   -   -   -   -   -   - 14
13. Stipules ciliolate; fig-bearing spurs usually up to c. 5 cm. long   -   -   -   -   -   -   -   -   -   - 32. *sansibarica* subsp. *sansibarica*
  - Stipules not ciliolate; fig-bearing spurs usually up to c. 10(15) cm. long   -   -   -   -   -   -   -   -   -   - 32. *sansibarica* subsp. *macrosperma*
14. Petiole (1)1.5–2 mm. thick when dry; peduncle 1–1.5 mm. thick when dry, if 0.5–1 mm. thick then leaf base acute to obtuse or the lamina 8–16 cm. long   -   -   -   -   -   -   - 28. *ottoniifolia* subsp. *macrosyce*
  - Petiole 0.5–1 mm. thick when dry; peduncle 0.5–1 mm. thick when dry; leaf base rounded to cordulate; lamina 2–9 cm. long   -   -   -   -   -   -   -   -   - 29. *tremula*
15. Stipules persistent or subpersistent   -   -   -   -   -   -   -   -   -   - 16
  - Stipules caducous   -   -   -   -   -   -   -   -   -   -   -   - 18
16. Stipules basally connate, 15–30 mm. long; figs at least 20 mm. in diam. when dry   -   -   -   -   -   -   -   -   -   -   -   - 35. *cyathistipula*
  - Stipules free, 5–15 mm. long; figs at most 10 mm. in diam. when dry   -   -   - 17
17. Petiole 2–10(15) mm. long; leaf glabrous, midrib not reaching leaf apex   -   -   -   -   -   -   -   -   -   -   -   -   - 24. *lingua*
  - Petiole 10–40(60) mm. long (if less than 10 mm. then leaf ± hairy), midrib reaching leaf apex   -   -   -   -   -   -   -   -   -   -   -   - 27. *thonningii*
18. Tertiary venation of leaf ± perpendicular to the lateral veins (scalariform, see TAB. 16, fig. 27); basal pair of veins strongly branched on the lower side; leaf base often cordate and/or the lower surface hairy   -   -   -   -   -   -   -   -   -   -   -   -   -   - 19
  - Tertiary venation of the leaf ± strongly reticulate or predominantly parallel to the lateral veins (TAB. 16, figs. 28, 29); basal pair of veins usually unbranched; leaf base usually cuneate to rounded and/or the lower surface glabrous   -   -   -   -   -   -   -   -   - 23
19. Lateral veins 3–7 pairs   -   -   -   -   -   -   -   -   -   - 14. *glumosa*
  - Lateral veins 7–16 pairs   -   -   -   -   -   -   -   -   -   -   - 20
20. Leaf ± as long as broad, cordiform to subreniform or broadly ovate; basal bracts caducous   -   -   -   -   -   -   -   -   -   -   -   - 19. *abutilifolia*
  - Leaf distinctly longer than broad, ovate to oblong or elliptic; basal bracts persistent   -   -   -   -   -   -   -   -   -   -   -   -   -   - 21
21. Leaf lateral veins 7–11 pairs; stipules 1.5–4.5(8) cm. long   -   -   - 20. *trichopoda*
  - Leaf lateral veins 10–16 pairs; stipules 0.5–1.2 cm. long   -   -   -   -   - 22
22. Peduncle 10–25 mm. long; conspicuous bud-cover absent; figs subglobose, if ellipsoid then up to 10 mm. in diam. when dry   -   -   -   -   -   -   - 12. *bussei*
  - Peduncle up to 5 mm. long; calyptrate bud-cover up to 15 mm. long (TAB. 16, fig. 21); figs ellipsoid, if subglobose then more than 10 mm. in diam. when dry   -   -   - 34. *ovata*

23. Basal bracts caducous - - - - - - - - - - - 24
 – Basal bracts persistent - - - - - - - - - - - 28
24. Figs 3–4 mm. in diam. when dry; leaf lamina 0.5–5 cm. long; petiole 2–8 mm.
     long - - - - - - - - - - - - - - - 24. *lingua*
 – Figs 5–20 mm. in diam. when dry; leaf lamina mostly more than 5 cm. long; petiole mostly more
     than 10 mm. long - - - - - - - - - - - - - 25
25. Leaf midrib not reaching apex of the lamina; apex bluntly and shortly acuminate to rounded or
     emarginate; lamina mostly 2.5–9 cm. long - - - - - - - 26
 – Leaf midrib reaching apex of the lamina; apex acuminate to acute; lamina mostly 10–20 cm.
     long - - - - - - - - - - - - - - - - 27
26. Figs 15–20 mm. in diam. when dry; leaf ovate to elliptic - - - - 22. *fischeri*
 – Figs 5–10 mm. in diam. when dry; leaf usually lanceolate to linear
     or oblong - - - - - - - - - - - - - 38. *barteri*
27. Figs 8–15 mm. in diam. when dry; leaves usually oblong to elliptic or subobovate; stipules
     glabrous or puberulous; basal bracts caducous - - - 25. *natalensis* subsp. *natalensis*
 – Figs 5–10 mm. in diam. when dry; leaves usually broadly obovate to obtriangular; stipules usually
     subsericeous or hirtellous; basal bracts caducous or
     subpersistent - - - - - - - - 25. *natalensis* subsp. *leprieurii*
28. Figs 10–20(25) mm. in diam. when dry - - - - - - - - - 29
 – Figs 3–12(15) mm. in diam. when dry - - - - - - - - - 31
29. Leaf lateral veins 3–7 pairs; basal veins usually branching (TAB. 16,
     fig. 25) - - - - - - - - - - - - - 14. *glumosa*
 – Leaf lateral veins (5)7–18 pairs; basal veins unbranched - - - - - 30
30. Petiole 2–3 mm. thick; leaf mostly 10–20 cm. long; figs 3–4.5 cm. in diam. when
     fresh - - - - - - - - - - - - - - 36. *scassellatii*
 – Petiole 1–2 mm. thick; leaf mostly 5–10 cm. long; figs at most 2 cm. in diam. when
     fresh - - - - - - - - - - - - - - 27. *thonningii*
31. Figs 3–4 mm. in diam. when dry; peduncle up to 5 mm. long; petiole 2–5 mm. long,
     0.5–1 mm. thick - - - - - - - - - - - - - 24. *lingua*
 – Figs 4–12(15) mm. in diam. when dry; peduncle more than 5 mm. long; petiole more than 8 mm.
     long and/or 1–2(2.5) mm. thick - - - - - - - - - 32
32. Leaf lateral veins (5)7–12(16) pairs - - - - - - - 27. *thonningii*
 – Leaf lateral veins 3–7 pairs - - - - - - - - - - 33
33. Twigs and leaves (especially below) usually hairy, if glabrous then leaves cordate; twigs 2–6 mm.
     thick; periderm on older parts of leafy twigs flaking off - - - - 14. *glumosa*
 – Twigs and leaves usually glabrous; twigs 1.5–3 mm. thick; periderm on older parts of leafy twigs
     not flaking off - - - - - - - - - - - - - 34
34. Leaf lateral veins (3)5–6(7) pairs; mature figs blackish or dark brown; plants from southern
     Mozambique - - - - - - - - - - - - 26. *burtt-davyi*
 – Leaf lateral veins (5)6–7 pairs; mature figs reddish or yellowish; plants of
     northwest Zambia - - - - - - - 25. *natalensis* subsp. *leprieurii*
35. Leaf glabrous beneath - - - - - - - - - - - - 36
 – Leaf hairy beneath - - - - - - - - - - - - - 44
36. Lateral veins 10–14 pairs - - - - - - - - - - - 37
 – Lateral veins 4–10 pairs - - - - - - - - - - - 38
37. Figs usually ellipsoid to ovoid, 10–40 mm. in diam. when dry; the calyptrate bud cover up to 15
     mm. long, at first enveloping the young fig (TAB. 16, fig. 21) - - - - 34. *ovata*
 – Figs subglobose, 5–10(15) mm. in diam. when dry; the calyptrate bud cover minute, at first ±
     enveloping the young fig, or not - - - - - - - - 27. *thonningii*
38. Epidermis of the petiole flaking off when dry - - - - - - - 39
 – Epidermis of the petiole not flaking off - - - - - - - - 40
39. Petiole 1–2 mm. thick when dry; basal bracts 1.5–2 mm.
     long - - - - - - - - - 37. *ardisioides* subsp. *camptoneura*
 – Petiole 2–4(8) mm. thick when dry; basal bracts 3–6 mm. long - - - - 21. *lutea*
40. Midrib not reaching the apex of the lamina - - - - - - - 41
 – Midrib reaching the apex of the lamina - - - - - - - - 42
41. Leaf apex often truncate to ± emarginate; leaves often subopposite; figs initially enveloped by a
     calyptrate bud-cover up to c. 10 mm. long (TAB. 16, fig. 21); the bud-cover halves and the
     stipules subpersistent - - - - - - - - - - 23. *craterostoma*
 – Leaf apex acuminate to rounded; leaves occasionally subopposite; figs initially enveloped by a
     minute bud-cover, or not so - - - - - - - - - 27. *thonningii*
42. Lateral veins (5)7–12(16) pairs - - - - - - - - 27. *thonningii*
 – Lateral veins 3–7 pairs - - - - - - - - - - - 43
43. Base of the lamina cordate (to rounded) - - - - - - 14. *glumosa*
 – Base of the lamina acute to obtuse - - - - 37. *ardisioides* subsp. *camptoneura*
44. Epidermis of the petiole flaking off when dry - - - - - - - 45
 – Epidermis of the petiole not flaking off - - - - - - - - 46

45. Leaf almost as long as wide; main basal lateral veins reaching the leaf margin near the middle of the lamina - - - - - - - - - - - - - - 13. *wakefieldii*
 – Leaf distinctly longer than wide; main basal lateral veins reaching the margin far below the middle of the lamina - - - - - - - - - - - - - 21. *lutea*
46. Main basal lateral veins (faintly) branched; base of the lamina usually subcordate - - - - - - - - - - - - - - - - 47
 – Main basal lateral veins unbranched (TAB. 16, fig. 26) - - - - - - 51
47. Reticulum of the lamina beneath inconspicuous and plane - - - 14. *glumosa*
 – Reticulum of the lamina beneath conspicuous and prominent - - - - 48
48. Leaf subreniform, cordiform or broadly ovate - - - - - - - 49
 – Leaf oblong, elliptic or ovate - - - - - - - - - - 50
49. Stipules 10–15 mm. long - - - - - - - - - - 18. *muelleriana*
 – Stipules 2–6 mm. long - - - - - - - - - - - 17. *tettensis*
50. Petiole 0.5–1 mm. thick when dry; figs 5–10 mm. in diam. when dry 16. *nigropunctata*
 – Petiole (1)2–3 mm. thick when dry; figs (7)10–15 mm. in diam. when dry 15. *stuhlmannii*
51. Reticulum of the leaf beneath prominent - - - - - - - 15. *stuhlmannii*
 – Reticulum of the leaf beneath plane or slightly prominent - - - - - 52
52. Lateral veins 3–7 pairs - - - - - - - - - - 14. *glumosa*
 – Lateral veins (5)7–12(16) pairs - - - - - - - - - 27. *thonningii*

1. **Ficus exasperata** Vahl, Enum. Pl. **2**: 197 (1805). —Hutch. in F.T.A. **6**, 2: 110 (1916). —Lebrun & Boutique in F.C.B. **1**: 126 (1948). —Keay in F.W.T.A. ed. 2, **1**: 605, fig. 173 (1958). —F. White, F.F.N.R.: 30 (1962). —K. Coates Palgrave, Trees Southern Africa: 107 (1977). —C.C. Berg et al. in Fl Cameroun **28**: 121, t. 39 (1985). —C.C. Berg in F.T.E.A., Moraceae: 52 (1989); in Kirkia **13**: 254 (1990). Type from Ghana.
 Ficus scabra Willd. in Mem. Acad. Berol. **1798**: 102 (1801) non Sim vide *F. sycomorus* (1909). Type from Guinea.
 Ficus punctifera Warb. in Ann. Mus. Congo, Bot., sér. 6, **1**: 35, t. 7 (1904). Lectotype from Zaire, chosen by Berg in Fl. Gabon **26**: 124 (1984).
 Ficus silicea Sim, For. Fl. Port. E. Afr.: 102, t. 87 (1909). Type: Mozambique, (Quissico), *Sim* 5381 (K, holotype).

Shrub or tree, up to 20(30) m. tall. Leafy twigs 1–5 mm. thick, hispidulous. Leaves almost distichous and alternate, sometimes subopposite; lamina ovate to elliptic or obovate, sometimes to oblong or subcircular, 2–16(20) × 1–12 cm., sometimes ± asymmetrical, subcoriaceous to coriaceous, apex shortly acuminate, sometimes acute, obtuse or rounded, base cuneate or occasionally subcordate, margin dentate to subentire; superior surface scabrous, hispidulous, the inferior surface scabrous, hispidulous or partly hirtellous to subtomentose; lateral veins 3–5(6) pairs; petiole 5–22(60) mm. long; stipules 2–5 mm. long, strigillose to strigose, caducous. Figs in pairs or solitary in the leaf axils, just below the leaves or sometimes on the older wood; peduncle 5–10(15) mm. long; bracts 1–5, broadly ovate, scattered on the peduncle, 1–4 similar bracts on the receptacle outer surface; receptacle subglobose, 10–25 mm. in diam. when fresh, 8–15 mm. in diam. when dry, hispidulous, yellow, orange or reddish at maturity.

Zambia. N: Lumangwe, 14.xi.1957, *Fanshawe* 4003 (FHO; K; NDO). **Zimbabwe**. E: Chipinge, Musirizwe/Bwazi R. junction area, 30.i.1975, *Pope, Biegel & Russell* 1458 (K; PRE). **Malawi**. N: Nkhata Bay, Mkuwadzi Forest, 14.xi.1982, *Dowsett-Lemaire* 18 (FHO). S: Thyolo Mt., 24.ix.1946, *Brass* 17792 (BM; K; PRE). **Mozambique**. N: Cabo Delgado, entre Mueda e Nangade, 18.ix.1948, *Barbosa* 2218 (K; LISC; PRE). Z: Milange, Serra Tumbine, 19.i.1966, *Correia* 487 (LISC). T: Monte de Zóbuè, 3.x.1942, *Mendonça* 568 (COI; LISC; LMU; SRGH). MS: Mossurize Distr., Mt. Espungabera, 21.xi.1960, *Leach & Chase* 10502 (BM; COI; K; LISC; PRE). GI: Quissico, 1908, *Sim* 5381 (K).
 From Senegal to Djibouti, extending to Angola in the southwest and Mozambique in the southeast. Also in São Tomé and Bioko, Yemen, Sri Lanka and S. India. Evergreen forests and forest margins, and riverine vegetation, often as a strangler, sometimes persisting in cleared places; 0–2000 m.

2. **Ficus asperifolia** Miq. in Hook., Lond. Journ. Bot. **7**: 564, t. 15B (1848); in Hook., Niger Fl.: 524 (1849). —Hutch. in F.T.A. **6**, 2: 111 (1916). —Lebrun & Boutique in F.C.B. **1**: 127 (1948). —Keay in F.W.T.A. ed. 2, **1**: 606 (1958). —F. White, F.F.N.R.: 29 (1962). —C.C. Berg et al. in Fl. Cameroun **28**: 124, t. 40 (1985). —C.C. Berg in F.T.E.A., Moraceae: 53 (1989); in Kirkia **13**: 254 (1990). Type from Nigeria.
 Ficus pendula Hiern, Cat. Afr. Pl. Welw. **1**, 4: 1008 (1900), non Link (1822). Type from Angola.
 Ficus urceolaris Hiern, tom. cit.: 1010 (1900). —Lebrun & Boutique in F.C.B. **1**: 128 (1948). Type from Angola.
 Ficus urceolaris var. *bumbana* Hiern, tom. cit.: 1010 (1900). Lectotype from Angola, chosen by Berg in Fl. Gabon **26**: 127 (1984).

Ficus storthophylla Warb., Ann. Mus. Congo, Bot., sér. 6, **1**: 32 (1904). —Lebrun & Boutique in
F.C.B. **1**: 128, t. 13 (1948). Type from Uganda.
Ficus scolopophora Warb. tom. cit.: 33 (1904). Type from Zaire.
Ficus xiphophora Warb. tom. cit.: 34, t. 10, 11 (1904). Type from Zaire.
Ficus irumuensis De Wild., Pl. Bequaert. **1**: 341 (1922). Type from Zaire.
Ficus storthophylla var. *cuneata* De Wild., Pl. Bequaert. **1**: 347 (1922). Type from Zaire.

Shrub up to 5 m. tall; branches often whippy, straggling or subscandent. Leafy twigs 1–5
mm. thick, white to brown hirtellous, hispidulous, ± strigillose or almost glabrous. Leaves
distichous, alternate; lamina elliptic to oblong, ovate, subobovate, lanceolate or
sometimes linear, 3–23 × 1.5–12 cm., usually ± asymmetrical, chartaceous to
subcoriaceous, apex acuminate to caudate, acute or sometimes obtuse, base cuneate or
rounded, margin dentate to irregularly pinnately lobed or divided, sometimes subentire;
superior surface scabrous, hispidulous or strigillose, the inferior surface hispid to
hirtellous, sometimes almost glabrous; lateral veins usually 3–10 pairs, up to 13 pairs in
large leaves, or more than 13 pairs in very narrow leaves; petiole 5–20 mm. long; stipules
3–6 mm. long, puberulous to almost glabrous, caducous. Figs 1–3(5) together in the leaf
axils or just below the leaves, sometimes on the older wood; peduncle (5)8–15 mm. long;
bracts 2–4, small, scattered on the peduncle, 2–4 similar bracts on the receptacle outer
surface; receptacle depressed globose to obovoid, c. 1.5–3 cm. in diam. when fresh, 1–1.5
cm. in diam. when dry, ± hispidulous, dark red to orange or yellowish at maturity.

Zambia. N: Kawambwa, 21.viii.1957, *Fanshawe* 3483 (K; NDO). W: West Lunga R., at Mwinilunga,
23.i.1975, *Brummitt, Chisumpa & Polhill* 14030 (BR; K; NDO).
    From Senegal to southern Sudan, Kenya, Uganda and Tanzania, westwards to Angola. Also on
Bioko. Evergreen forest and forest margins, riverine vegetation, and wooded grassland; 0–1800 m.
    This species is very variable with two forms recognisable: an eastern form in which the figs are
short-pedunculate and a western form in which the figs are long-pedunculate. It may be
distinguished from *F. capreifolia* and *F. pygmaea* by the leaves which in these two species are more
symmetrical at the base.

3. **Ficus pygmaea** Welw. ex Hiern, Cat. Afr. Pl. Welw. **1**, 4: 1009 (1900). —Hutch. in F.T.A. **6**, 2: 106
    (1916). —F. White, F.F.N.R.: 29 (1962). —C.C. Berg in Kirkia **13**: 255 (1990). —van Greuning in S.
    Afr. J. Bot. **56**, 6: 601, fig. 2 (1990). TAB. **17**. Type from Angola.
    Ficus brevicula Hiern, Cat. Afr. Pl. Welw. **1**, 4: 1009 (1900). Type from Angola.

Shrub up to 1.5(4) m. tall. Leafy twigs 1.5–4 mm. thick, hirtellous to puberulous. Leaves
spiral to distichous, alternate; lamina oblong to elliptic or subovate to lanceolate, or
sublinear, 1–13.5 × 0.5–6 cm., somewhat asymmetrical, chartaceous, apex acute
sometimes obtuse, occasionally tridentate, base cuneate rounded or subcordate, margin
remotely serrate to deeply and irregularly lobed, sometimes subentire; rough on both
surfaces, scabrous and hispidulous above, hirtellous to pubescent below; lateral veins
(3)5–9 pairs; petiole 1–8(12) mm. long, 0.5–2 mm. thick; stipules 2–6 mm. long,
puberulous, caducous, or subpersistent. Figs solitary, axillary; peduncle 5–20 mm. long;
bracts 1–3, small, scattered over the peduncle, sometimes whorled; receptacle
subpyriform to ellipsoid or subglobose, up to 3 cm. in diam. when fresh, (0.5)1–1.5 cm. in
diam. when dry, hispidulous, yellow, pink, red or purple at maturity.

Caprivi Strip: Singalambwe, c. 1100 m., 31.xii.1958, *Killick & Leistner* 3213 (K; PRE).
Botswana. N: Ngamiland, Okavango R., near Sepopa, 6.v.1975, *Müller & Biegel* 2324 (K; PRE).
Zambia. B: Ngonye (Gonye) Falls, Zambezi R., 27.vii.1952, *Codd* 7208 (K; PRE; SRGH). N: Mansa to
Kawambwa, "Mile 52", 29.x.1952, *White* 3521 (BM; COI; FHO). W: Mwinilunga, W. of Kalene Hill
Mission, 22.ix.1952, *White* 3326 (BM; FHO; PRE). S: Katambora, riverside, 20.i.1956, *Gilges* 568 (K;
PRE).
    Also in Namibia and Angola. Flood plains and moist sandy soil near rivers and streams often
forming thickets.
    This species in confined almost entirely to the Kalahari Sands, with little or no overlap with the
area of *F. capreifolia*.

4. **Ficus capreifolia** Del. in Ann. Sci. Nat., Bot., sér. 2, **20**: 94 (1843) as "*capreaefolia*". —Warb. in Ann.
    Mus. Congo, Bot., sér. 6, **1**: 36, t. 22 (1904). —Mildbr. & Burret in Engl., Bot. Jahrb. **46**: 202 (1911).
    —Hutch. in F.T.A. **6**, 2: 107 (1916); in F.C. **5**, 2: 531 (1925). —Lebrun & Boutique in F.C.B. **1**: 126
    (1948). —Keay in F.W.T.A. ed. 2, **1**: 605 (1958). —F. White in F.F.N.R. : 29, t. 6A (1962). —K.
    Coates Palgrave, Trees Southern Africa: 105 (1977). —C.C. Berg et al. in Fl. Cameroun **28**: 127, t.
    41 (1985). —C.C. Berg in F.T.E.A., Moraceae: 53, t. 17 (1989); in Kirkia **13**: 255 (1990). —van
    Greuning in S. Afr. J. Bot. **56**, 6: 601, fig. 1 (1990). Type from Ethiopia.

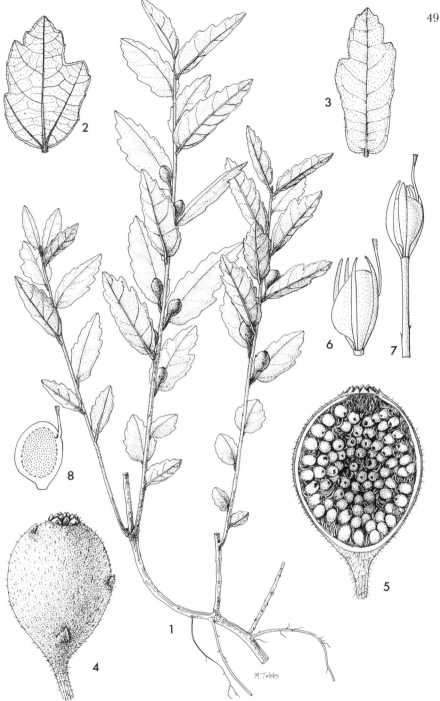

Tab. 17. FICUS PYGMAEA. 1, habit (×⅔); 2 & 3, leaves (× 1); 4, fig (× 4); 5, l/s through fig (× 4), 1–5 *Welwitsch* 6370; 6, diagram of sessile long-styled pistillate flower (after anthesis) with 4(6) free tepals and long style (× 15); 7, diagram of pedicellate long-styled pistillate flower (after anthesis) with tepals and short style (× 15); 8, diagramatic l/s through fruit showing endocarp and exocarp (× 15), 6–8 *Pope* 339. Drawn by M. Tebbs & S. Dawson.

Ficus antithetophylla Miq. in Hook., Lond. Journ. Bot. **7**: 236, t. 5B (1848). Type from Ethiopia.
Ficus palustris Sim, For. Fl. Port. E. Afr.: 99, t. 90C (1909). Type: Mozambique, Mucabilla R., Maganja (Magenga) da Costa, *Sim* 6237 (K, lectotype, chosen here by Berg).

Shrub up to 6 m. tall. Stems slender, usually with stiff branches. Leafy twigs 1–5 mm. thick, puberulous to hirsute. Leaves ± distinctly distichous and alternate or subopposite to subverticillate; lamina subovate to oblong or lanceolate, sometimes elliptic, 2–15 × 1–5.5 cm., symmetrical, chartaceous, apex acute, obtuse or rounded, 3-lobed or 3-dentate, base rounded to obtuse or cuneate, margin subentire or ± faintly crenate; both surfaces scabrous; lateral veins 5–12 pairs; petiole 2–10(25) mm. long; stipules 5–10 mm. long, partly puberulous, mostly subpersistent. Figs solitary or in pairs in the leaf axils; peduncle 5–15 mm. long with 3 small bracts scattered or whorled; receptacle stipitate, without bracts on the outer surface, 1.5–3 cm. in diam. when fresh, globose, often pyriform and 1–2.5 cm. in diam. when dry, hispidulous, green to pale yellow at maturity, stipe up to c. 5 mm. long.

Botswana. N: Moanachira R., 9.v.1973, *P.A. Smith* 602 (LISC; SRGH). **Zambia**. B: Ngonye (Gonya) Falls, 27.vii.1952, *Codd* 208 (BM). N: Mporokoso Distr., 28.x.1949, *Hoyle* 1327 (FHO). W: Kifubwa R. Gorge c. 5 km. S. of Solwezi, 26.xi.1982, *Berg & Bingham* 1427 (K; U). E: Chikowa Mission, 13.x.1958, *Robson* 77 (BM; K; LISC; PRE). S: Kafue R., Guabe Camp, 17.xi.1982, *Berg & Bingham* 1378 (K; U). **Zimbabwe**. N: Makonde Distr., Gomberra Ranch, Manyame (Hunyani) R., 18.viii.1968, *Jacobsen* 3466 (PRE). W: Zambezi R., near confluence with Deka R., 16.xii.1973, *Raymond* 235 (BM; SRGH). C: Chegutu Distr., Umniati R., near Sanyati Reserve Rest House, 13.xii.1962, *Müller* 6 (BM; K; LISC; SRGH). E: Mutare Distr., Nyamkwarara Valley, Stapleford, 1.xi.1967, *Mavi* 370 (K; LISC; SRGH). S: Runde (Lundi) R., near Mozambique border, 20.x.1946, *Hornby* 2496 (K; PRE). **Malawi**. C: Chitala, Doma, 29.x.1941, *Greenway* 6386 (K). S: Island in the Shire R., below Mbewe, x.1887, *Scott* s.n. (K). **Mozambique**. N: Ribáuè, R. Meperipui, 21.i.1964, *Torre & Paiva* 10072 (LISC; LMA; UC). Z: Entre Luabo e Chinde, 13.x.1941, *Torre* 3636 (LISC). T: Arredores de Lobuè, 17.vi.1941, *Torre* 2882 (LISC). MS: Lamego (Vila Machado), R. Machire (M'Tuchira), 17.iv.1948, *Mendonça* 4009 (COI; LISC; LMU). GI: Guijá Distr., Massingir, 2.xii.1944, *Mendonça* 3225 (LISC; PRE; SRGH). M: Arredores de Magude, 16.ii.1948, *Torre* 7152 (K; LISC; WAG).

From Senegal to the Somali Republic and southwards to northern Namibia and to Natal in South Africa. Locally common on banks and floodplains of low altitude rivers, on sand banks and swampy places, usually forming dense thickets; 0–1600 m.

5. **Ficus sycomorus** L., Sp. Pl.: 1059 (1753). —Mildbr. & Burret in Engl., Bot. Jahrb. **46**: 191 (1911). —Hutch. in F.T.A. **6**, 2: 95 (1916); in F.C. **5**, 2: 526 (1925). —F. White, F.F.N.R.: 30, t. 6B, 6J (1962). —Gomes e Sousa, Dendrol. Moçamb. Estudo Geral **1**: 190, t. 10 (1967). —K. Coates Palgrave, Trees Southern Africa: 116 (1977). —C.C. Berg in Kirkia **13**: 256 (1990). —van Greuning in S. Afr. J. Bot. **56**, 6: 602, fig. 3 (1990). TAB. **18**. Type from Egypt, not found (not in LINN).
    Sycomorus antiquorum Gasp., Ricerche Caprifico: 56 (1845). Type as above.
    Sycomorus gnaphalocarpa Miq. in Hook., Lond. Journ. Bot. **7**: 113 (1848). Type from Ethiopia.
    Ficus damarensis Engl., Bot. Jahrb. **10**: 5 (1888). Type from Namibia.
    Ficus integrifolia Sim, For. Fl. Port. E. Afr.: 101, t. 89 (1909). Type: Mozambique, *Sim* 6145 (not traced).
    Ficus scabra Sim, For. Fl. Port. E. Afr.: 102, t. 95c (1909) non Foster f. (1786) nec Willd. (1801). Type: Mozambique, Nhamacurra, *Sim* 5644 (K, holotype).

Tree up to 20(30) m. tall, trunk short, up to c. 2 m. diam., main branches spreading. Leafy twigs 2–6 mm. thick, densely, minutely puberulous and with much longer, white to yellowish hairs, especially on the nodes; periderm flaking off when dry. Leaf lamina ovate to elliptic, or obovate to subcircular (1)2.5–12(21) × (0.5)2–11(16) cm., chartaceous to coriaceous, apex rounded to obtuse, base cordate sometimes obtuse, margin subentire, slightly repand or denticulate; superior surface scabrous to scabridulous, sometimes almost smooth or hispidulous to strigillose, the main veins whitish hirtellous to hirsute, inferior surface puberulous to hispidulous, the main veins sparsely whitish hirtellous to hirsute; lateral veins 5–10 pairs; petiole (0.5)1–4(6) cm. long, 1–3 mm. thick, densely, minutely white puberulous and with much longer white to yellowish hairs, with the periderm flaking off when dry; stipules 5–25 mm. long, white puberulous to tomentose or partly hirtellous to hirsute, caducous. Figs solitary or sometimes in pairs in the leaf axils or just below the leaves on unbranched, leafless branchlets up to 20 cm. long, or on branched, leafless branchlets up to 20(35) cm. long, or borne on the older branches down to the trunk; peduncle 3–25 mm. long, 1–3 mm. thick; basal bracts 2–3 mm. long. Receptacle obovoid to pyriform or subglobose, often stipitate, at least when dry, 1.5–5 cm. in diam. when fresh, (1)1.5–3 cm. in diam. when dry, velutinous or densely tomentose to sparsely puberulous or pubescent, sometimes almost glabrous, yellowish to reddish at maturity.

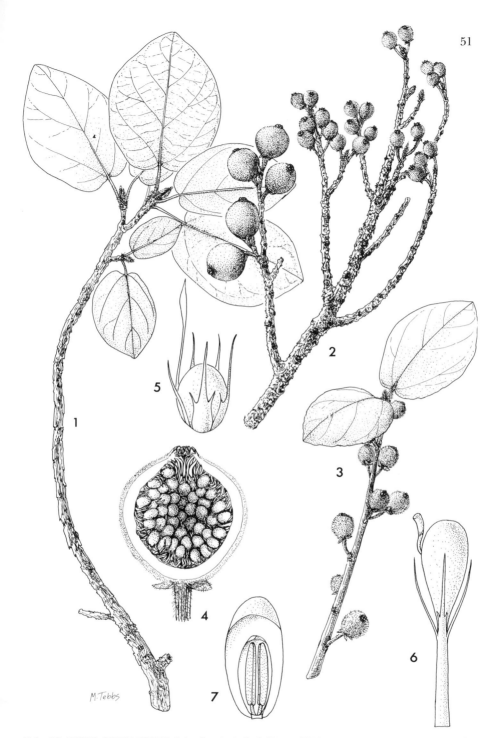

Tab. 18. FICUS SYCOMORUS. 1, leafy twig (× ⅔), *de Koning* 7758; 2, fertile twig (× ⅔), *de Koning &*
Boane 8670; 3, fertile twig (× ⅔), *Excell & Mendonça* 2863; 4, l/s through fig (× 2), *Müller* 4; 5,
long-styled pistillate flower (× 12); 6, short-styled pistillate flower (× 14); 7, staminate flower (2
stamens enveloped by 2 bracteoles and a cucullate outer perianth with overlapping lobes (× 14),
5–7 *Breteler* 319. Drawn by M. Tebbs & S. Dawson.

Caprivi Strip. Linyanti Swamp near Shaile (map sq. 1824), 28.x.1972, *Biegel, Pope & Russell* 4092 (K; SRGH). **Botswana**. N: Okavango Swamps, near Maun, Thamalakane R. bank, 25.iv.1971, *P.A. Smith* 56 (K; SRGH). **Zambia**. B: Mwande (Old Sesheke), near Zambezi R., 25.xii.1952, *Angus* 1026 (BM; FHO; K; PRE). N: Lake Bangweulu, N. of Samfya Mission, 22.viii.1952, *Angus* 288 (K; PRE). W: Mwinilunga Distr., near Matonchi R., above dam, 24.x.1937, *Milne-Redhead* 2932 (BM; K; PRE). C: c. 9 km. E. of Lusaka, 4.iii.1952, *White* 2185 (FHO; K). S: near Mr. Watts Farm, c. 40 km. NE. of Choma, 5.vi.1952, *White* 2922 (FHO; K). **Zimbabwe**. N: Makonde, Chirombodzi Farm, 29.ix.1968, *Jacobsen* 3525 (PRE). W: Bulawayo, Kumalo, 14.viii.1954, *Orpen* 50/54 (K; SRGH). C: Chegutu, Umniati R., near Sanyati Reserve Rest House, 13.xii.1962, *Müller* 4 (BM; K; LISC; PRE). E: Mutare, Umtali Boys High School, 28.vii.1962, *Chase* 7788 (K; SRGH). S: Gwanda Distr., Bubye R., Ranch Homestead, 3.v.1958, *Drummond* 5534 (K; LISC; PRE). **Malawi**. N: Nkhata Bay Distr., at Kilwa, c. 92 km. S. of Nkhata Bay Junction, 27.viii.1976, *Pawek* 11732 (K; LMA; MAL; SRGH; UC). C: Nkhota Kota, Chia, 7.ix.1946, *Brass* 17566 (K). S: Mulanje Mt., 14.viii.1957, *Chapman* 408 (FHO; K). **Mozambique**. N: Mutuali, near the bridge of Nalume R., 21.ix.1953, *Gomes e Sousa* 4124 (COI; K; PRE). Z: Maganja da Costa, between Raraga R. and Muzo, 28.ix.1949, *Barbosa & Carvalho* 4245 (K). T: Mágoè Distr., 17 km. from Mágoè Velho to Zumbo, 2.iii.1970, *Torre & Correia* 18162 (LISC; LMA; LMU). MS: Chimoio Distr., between Zembe and R. Revuè, 30.iv.1948, *Andrada* 1212 (PRE). GI: Nhanvue, c. 60 km. sud de Inhambane, xi.1935, *Gomes e Sousa* 1680 (K). M: entre Boane e Porto Henrique, 19.ii.1981, *de Koning & Boane* 8670 (K; LMU).

Widespread in subsaharan Africa, extending to the Arabian Peninsula in the northeast and to Namibia and South Africa (Natal) in the south. Also on the Comoro Islands, Cape Verde Islands and Madagascar. Wooded grassland and important in riverine fringes; 0–2000 m.

Cultivated for its fruit in Middle East countries.

6. **Ficus sur** Forssk., Fl. Aegypt.-Arab.: CXXIV, 180 (1775). —Hutch. in F.T.A. **6**, 2: 100 (1916). —Aweke in Meded. Landb. Wag. **79**-3: 66, t. 17 (1979). —C.C. Berg et al. in Fl. Cameroun **28**: 135, t. 44 (1985). —C.C. Berg in F.T.E.A., Moraceae: 56, fig. 18 (1989); in Kirkia **13**: 256 (1990). —van Greuning in S. Afr. J. Bot. **56**, 6: 604, fig. 4 (1990). Type from Yemen.

Ficus capensis Thunb., Diss. Fic.: 13 (1786). —Mildbr. & Burret in Engl., Bot. Jahrb. **46**: 195 (1911). —Hutch. in F.T.A. **6**, 2: 101 (1916); in F.C. **5**, 2: 527 (1925). —Lebrun & Boutique in F.C.B. **1**: 116 (1948). —Keay in F.W.T.A. ed. 2, **1**: 606 (1958). —F. White, F.F.N.R.: 30, t. 6C (1962). —K. Coates Palgrave, Trees Southern Africa: 105 (1977). Type from S. Africa (not found in UPS —THUNB).

Ficus lichtensteinii Link, Enum. Hort. Berol. **2**: 451 (1822). Type from S. Africa.

Sycomorus thonningiana Miq. in Hook., Lond. Journ. Bot. **7**: 112, 563, t. 14A (1848) nomen; in Verh. Eerste Kl. Kon. Ned. Inst. Wet. Amsterdam, ser. 3, **1**: 123 (1849). Lectotype from the Cape Verde Islands, chosen by Berg in Fl. Gabon **26**: 135 (1984).

Sycomorus capensis (Thunb.) Miq. in Hook., Lond. Journ. Bot. **7**: 113, t. 3B (1848). Type as for *Ficus capensis*.

Sycomorus sur (Forssk.) Miq. in Verh. Eerste Kl. Kon. Ned. Inst. Wet. Amsterdam, ser. 3, **1**: 121 (1849). Type as for *Ficus sur*.

Ficus capensis var. *trichoneura* Warb. in Engl., Bot. Jahrb. **20**: 153 (1894). Type from Tanzania.

Ficus mallotocarpa Warb. in tom. cit.: 154 (1894). Type from Tanzania.

Ficus clethrophylla Hiern, Cat. Afr. Pl. Welw. **1**, 4: 1017 (1900). Lectotype from Angola, chosen by Berg in Fl. Gabon **26**: 135 (1984).

Ficus sycomorus var. *prodigiosa* Hiern, tom. cit.: 1012 (1900). Lectotype from Angola, chosen by Berg in Fl. Gabon **26**: 135 (1984).

Ficus sycomorus var. *alnea* Hiern, tom. cit.: 1013 (1900). Lectotype from Angola, chosen by Berg in Fl. Gabon **26**: 135 (1984).

Ficus sycomorus var. *polybotrya* Hiern, tom. cit.: 1014 (1900). Type from Angola.

Ficus capensis var. *pubescens* Warb. ex De Wild. & T. Durand in Ann. Mus. Congo, Bot. sér. 3, **2**: 215 (1901). Type from Zaire.

Ficus kondeensis Warb. in Engl., Bot. Jahrb. **30**: 292 (1901). Type from Tanzania.

Ficus plateiocarpa Warb. in loc. cit. (1901). Type from Tanzania.

Ficus stellulata var. *glabrescens* Warb. in Ann. Mus. Congo, Bot. sér. 6, **1**: 27 (1904). Lectotype from Zaire, chosen here by Berg.

Ficus villosipes Warb. in Ann. Mus. Congo, Bot. sér. 6, **1**: 28 (1904). Type from Zaire.

Ficus munsae Warb. in Ann. Mus. Congo, Bot. sér. 6, **1**: 29, t. 17 (1904). Type from Zaire.

Ficus erubescens Warb. in Ann. Mus. Congo, Bot. sér. 6, **1**: 29, t. 6 (1904). Type from Zaire.

Ficus capensis var. *mallotocarpa* (Warb.) Mildbr. & Burret in Engl., Bot. Jahrb. **46**: 198 (1911).

Ficus gongoensis De Wild. in Fedde, Repert. **12**: 196 (1913). Type from Zaire.

Ficus ostiolata De Wild. in Bull. Soc. Roy. Bot. Belg. **52**: 220 (1914). Type from Zaire.

Ficus ostiolata var. *brevipedunculata* De Wild. in Bull. Soc. Roy. Bot. Belge. **52**: 221 (1914). Type from Zaire.

Ficus beniensis De Wild. in Ann. Soc. Sci. Brux. **40**: 278 (1921). Type from Zaire.

Ficus ituriensis De Wild. in Ann. Soc. Sci. Brux. **40**: 281 (1921). Type from Zaire.

Tree up to 25(30) m. tall. Leafy twigs 2–5 mm. thick, white to yellowish (or brownish) puberulous, hirtellous, tomentose or hirsute to glabrescent, periderm usually not flaking off when dry. Leaf lamina elliptic to ± ovate or oblong, sometimes subcircular or lanceolate, 4–20(32) × 3–13(16) cm., chartaceous to coriaceous, apex acuminate to acute, base subacute to cordate, margin coarsely crenate-dentate to repand or ± entire; superior surface smooth, sometimes scabrous, glabrous or puberulous on the proximal parts on the main veins, the inferior surface puberulous to tomentose, or glabrous and only the main veins pubescent; lateral veins (3)5–9 pairs; petiole 1.5–9 cm. long, 1–2 mm. thick, puberulous, ± hirsute or subtomentose sometimes glabrous, epidermis usually not flaking off when dry; stipules 1–3.5 cm. long, white to yellowish subsericeous, pubescent or hirsute to glabrescent, caducous. Figs on branching, leafless branchlets up to 50(150) cm. long, on the older wood, down to the trunk, or figs occasionally in the leaf axils or just below the leaves; peduncle 5–20 mm. long, 1–3 mm. thick; basal bracts 2–3 mm. long. Receptacle obovoid to subglobose, often ± depressed-globose, often stipitate, at least when dry, 2–4 cm. in diam. when fresh, 0.5–2.5 cm. in diam. when dry, indumentum white to yellowish, ± puberulous to almost glabrous or densely tomentose to subvelutinous, red to dark orange at maturity.

Botswana. SE: Palapye Ratholo, along Tswapong Hills, 5.iii.1957, *de Beer* T2 (K; SRGH). **Zambia**. B: near Shangombo, Mashi R., 16.viii.1952, *Codd* 7569 (K; PRE). N: Mbala Distr., Kawimbe, Lumi R., 1680 m., 12.x.1956, *Richards* 6427 (K). W: Mwinilunga Distr., Matonchi R., 11.xi.1937, *Milne-Redhead* 3187 (BM; K; PRE). C: c. 10 km. E. of Lusaka, 4.iii.1952, *White* 2184 (FHO; K). E: Nyimba to Minga, 26.viii.1929, *Burtt Davy* 2094B (FHO). S: Zambezi R. at Katambora, 26.viii.1947, *Brenan & Greenway* 7760 (FHO; K). **Zimbabwe**. N: Near Mazowe (Mazoe) Dam Wall, 27.ix.1960, *Rutherford-Smith* 195 (K; PRE). W: Matopos Reserve, Korakora Stream near Absent Farm, 7.ix.1952, *Plowes* 1473 (PRE). C: Gweru Distr., Mlezu School Farm, c. 29 km. SSE. of Kwekwe, 5.i.1966, *Biegel* 1180 (K; SRGH). E: Mutare Distr., E. of Vumba Mts., 23.x.1967, *Chase* 8468 (K; LISC; SRGH). S: Mberengwa Distr., Buhwa Mt., c. 1100 m., 30.iv.1973, *Pope* 1022 (K; SRGH). **Malawi**. N: Viphya Plateau, 13.iii.1962, *Chapman* 1619 (K; SRGH). C: Ntchisi Mt., 1400 m., 30.vii.1946, *Brass* 17033 (K; PRE). S: Thyolo (Cholo) Mt., 24.ix.1946, *Brass* 17775 (K; PRE). **Mozambique**. N: Niassa Distr., Mutuali Region near Catholic Mission, 6.x.1953, *Gomes e Sousa* 4139 (COI; K; PRE). Z: Gúruè, 29.vi.1943, *Torre* 5616 (C; LISC; LMA). T: Monte Zóbuè, 3.x.1942, *Mendonça* 632 (EA; LISC; SRGH; WAG). MS: Manica Distr., Mavita, Monte Xiroso, 26.x.1944, *Mendonça* 2637 (K; LISC; WAG). GI: Chongoene, 24.x.1947, *Barbosa* 504 (COI). M: estrada para Goba Fronteira, próx. da fonte J Paco, 28.vi.1961, *Dulsinhus* 495 (K).

Widely distributed in subsaharan Africa, from Cape Verde to the Yemen and south to Angola and S. Africa (Cape Province). Also on the islands of São Tomé, Príncipe and Bioko. Widespread at most altitudes in wooded grassland, riverine fringes and evergreen forest; 0–2300 m.

This species is very variable in the indumentum of the leaves and figs, from glabrous to densely pubescent, and in leaf shape and lamina margin.

7. **Ficus vallis-choudae** Del. in Ann. Sci. Nat., Bot. sér. 2, **20**: 94 (1843). —Hutch. in F.T.A. **6**, 2: 103 (1916). —Lebrun & Boutique in F.C.B. **1**: 119 (1948). —Keay in F.W.T.A. ed. 2, **1**: 606 (1958). —F. White, F.F.N.R.: 30 (1962). —K. Coates Palgrave, Trees Southern Africa: 118 (1977). —C.C. Berg et al. in Fl. Cameroun **28**: 142, t. 47 (1985). —C.C. Berg in F.T.E.A., Moraceae: 58 (1989); in Kirkia **13**: 257 (1990). Type from Ethiopia.

    Sycomorus schimperiana Miq. in Hook., Lond. Journ. Bot. **7**: 112 (1848). Type as for *F. vallis-choudae.*

    Ficus schweinfurthii Miq. in Ann. Mus. Bot. Lugd.-Bat. **3**: 295 (1867). Type from Sudan.

    Ficus vallis-choudae var. *pubescens* Peter in Fedde, Repert., Beih. **40**, 2: 98 (1932). Types from Tanzania.

Tree up to 10(20) m. tall. Leafy twigs 2–10 mm. thick, glabrous or sparsely appressed puberulous, sometimes hirtellous or tomentose, indumentum white, periderm flaking off when dry. Leaf lamina ovate to cordiform or ± deltate, 4–24(36) × 3–24(30) cm., coriaceous to subcoriaceous, apex acute to subobtuse or very shortly acuminate, base obtuse to truncate or cordate, margin coarsely and obtusely dentate to repand, sometimes subentire; superior surface smooth and glabrous or puberulous on the main veins, sometimes scabridulous and hirtellous to pubescent; lateral veins 5–8 pairs; petiole 2–11(13.5) cm. long, 1–3 mm. thick, glabrous, appressed puberulous or sometimes hirtellous to tomentose, indumentum white, epidermis flaking off when dry; stipules 1–3 cm. long, ciliolate in the lower part, appressed puberulous or subsericeous, caducous. Figs solitary in the leaf axils or just below the leaves, occasionally on leafless branchlets up to 30 cm. long on the older wood; peduncle 2–12 mm. long, 4–6 mm. thick; basal bracts c. 2 mm. long. Receptacle subglobose to obovoid, c. 20 mm. in diam. when fresh, 15 mm. in diam. when dry, ± densely white to yellowish puberulous to hirtellous or tomentose, tomentellous near the ostiole, or glabrous; yellowish to orange at maturity with longitudinal orange to reddish stripes.

Zambia. N: Mkupa near hot spring, 7.x.1949, *Bullock* 1165 (K). W: Lake Kashiba, NW. of Mpongwe, 23.xi.1982, *Berg & Bingham* 1398 (K; U). C: Serenje Distr., Muchinga, Manda Stream, 30.xi., *J.C.M.* 3166 (FHO; NDO). **Zimbabwe**. E: Chimanimani Distr., Lusitu R., 8.viii.1975, *Müller* 2355 (PRE; SRGH). **Malawi**. N: Rumphi Distr., Livingstonia Escarpment, 14.i.1975, *Pawek* 9746 (K; PRE; SRGH). S: Thyolo (Cholo) Mt., 29.ix.1946, *Brass* 17875 (BM; K; PRE; SRGH). **Mozambique**. N: entre Maniamba e Metangula, 10.x.1942, *Mendonça* 730 (COI; LISC; PRE; SRGH). Z: Tumbini Hill, 19.ix.1942, *Hornby* 2809 (PRE).

Widespread in subsaharan Africa, from Guinea and Mali to Ethiopia and southwards to the Flora Zambesiaca area. Riverine forest, mixed evergreen forest and swamp forest (mushitu); 0–1800 m.

Fanshawe 1493 (K) from Chingola, Zambia, differs from typical *F. vallis-choudae* in that it bears figs on branching, leafless branchlets up to 30 cm. long, on the older wood.

This is the only non-cauliflorous continental species of subgen. Sycomorus. It can be distinguished from ramiflorous forms of *F. sycomorus* by the smooth leaf lamina.

8. **Ficus dicranostyla** Mildbr. in Engl., Bot. Jahrb. **46**: 204 (1911). —Hutch. in F.T.A. **6**, 2: 119 (1916). —Keay in F.W.T.A. ed. 2, **1**: 607 (1958). —F. White, F.F.N.R.: 30 (1962). —Aweke in Meded. Landb. Wag. **79**-3: 18, fig. 4 (1979). —C.C. Berg et al. in Fl. Cameroun **28**: 146, t. 48 (1985). —C.C. Berg in F.T.E.A., Moraceae: 59, t. 19 (1989); in Kirkia **13**: 258 (1990). Lectotype from Guinea, chosen by Berg in Fl. Gabon **26**: 144 (1984).

Ficus dicranostyla var. *nitida* Hutch. in F.T.A. **6**, 2: 120 (1916). Lectotype from Central African Republic, chosen by Berg in Fl. Gabon **26**: 144 (1984).

Ficus lynesii Lebrun in Bull. Séance. Inst. Roy. Col. Belge. **6**: 494 (1935). —Lebrun & Boutique in F.C.B. **1**: 124 (1948). Type from Zaire.

Tree up to 6(20) m. tall, or a shrub. Leafy twigs 2–5 mm. thick, white pubescent to puberulous. Leaf lamina elliptic to oblong or subovate, (2)5–20 × (1)2–9 cm., subcoriaceous to chartaceous, apex acuminate, sometimes subacute, base cuneate to subcordate, margin subentire; superior surface scabrous, the main veins puberulous to hirtellous or pubescent, inferior surface puberulous to hirtellous or pubescent; lateral veins 5–8(9) pairs; petiole 1–3.5 cm. long, 1–2 mm. thick; stipules 5–15 mm. long, puberulous to pubescent, caducous. Figs solitary or in pairs in the leaf axils; peduncle 3–10 mm. long; basal bracts 2–2.5 mm. long, persistent. Receptacle often shortly stipitate, at least when dry, globose to obovoid, 1–2.5 cm. in diam. when fresh, 0.5–1.5 cm. in diam. when dry, puberulous to hispidulous, yellowish to pale orange at maturity. Wall of fruiting fig c. 1.5 mm. thick when dry, firm.

Zambia. N: Kawambwa, *Bonds* 829 (NDO). W: Kitwe, 25.iii.1955, *Fanshawe* 2227 (K; LISC; NDO); Muchili, c. 53 km. from Ndola on road to Kapiri Mposhi, 27.xi.1982, *Berg & Bingham* 1436 (K; U).

Also from Senegal to Ethiopia, NE. and S. Zaire (Shaba) and Uganda. Miombo woodland, often on termite mounds in Zambia and Shaba Province of Zaire; 0–1300 m.

9. **Ficus ingens** (Miq.) Miq. in Ann. Mus. Bot. Lugd.-Bat. **3**: 288 (1867). —Hutch. in F.T.A. **6**, 2: 121 (1916); in F.C. **5**, 2: 529 (1925). —Lebrun & Boutique in F.C.B. **1**: 121 (1948). —Keay in F.W.T.A. ed. 2, **1**: 607 (1958). —F. White, F.F.N.R.: 30, t.7 (1962). —K. Coates Palgrave, Trees Southern Africa: 109 (1977). —C.C. Berg et al. in Fl. Cameroun **28**: 149, t. 49 (1985). —C.C. Berg in F.T.E.A., Moraceae: 60, t. 20 (1989); in Kirkia **13**: 259 (1990). —van Greuning in S. Afr. J. Bot. **56**, 6: 607, fig. 7 (1990). Type from Ethiopia.

Urostigma ingens Miq. in Hook., Lond. Journ. Bot. **6**: 554 (1847). Type from Ethiopia.

Urostigma caffrum Miq. in Verh. Eerste Kl. Kon. Ned. Inst. Wet. Amsterdam, ser. 3, **1**: 141 (1849). Type from S. Africa.

Urostigma xanthophyllum var. *ovatocordatum* Sond. in Linnaea **23**: 136 (1850). Type from S. Africa.

Ficus schimperiana A. Rich., Tent. Fl. Abyss. **2**: 266 (1851). Type as for *U. ingens*.

Ficus caffra (Miq.) Miq. in Ann. Mus. Bot. Lugd.-Bat. **3**: 288 (1867). Type from S. Africa.

Ficus stuhlmannii var. *glabrifolia* Warb. in Engl., Bot. Jahrb. **20**: 162 (1894). Type from Tanzania.

Ficus caffra var. *sambesiaca* Warb. in Viert. Nat. Ges. Zürich **51**: 140 (1906). Type: Mozambique, Boruma, Sambesi Mittellauf, *Menyhart* 770 (Z, holotype).

Ficus caffra var. *longipes* Warb. in Viert. Nat. Ges. Zürich **51**: 140 (1906). Type from S. Africa.

Ficus caffra var. *natalensis* Warb. in Viert. Nat. Ges. Zürich **51**: 140 (1906). Type from S. Africa.

Ficus caffra var. *pubicarpa* Warb. in Viert. Nat. Ges. Zürich **51**: 140 (1906). Type from S. Africa.

Ficus pondoensis Warb. in Viert. Nat. Ges. Zürich **51**: 140 (1906). Lectotype from S. Africa, chosen by Berg in Fl. Gabon **26**: 148 (1984).

Ficus magenjensis Sim, For. Fl. Port. E. Afr.: 99, t. 93B (1909). Type: Mozambique, Maganja da Costa, *Sim* 5653 (K, holotype).

Ficus ovaticordata De Wild. in Ann. Soc. Sci. Brux. **40**: 281 (1921). Type from Zaire.

Ficus ingens var. *tomentosa* Hutch. in F.C. **5**, 2: 530 (1925); in F.W.T.A. ed. 2, **1**: 607 (1958). Type from S. Africa.

Tree up to 15 m. tall. Leafy twigs 3–6 mm. thick, sparsely pubescent to densely tomentellous or subvelutinous, indumentum white to brownish. Leaf lamina ovate to elliptic, oblong or sometimes lanceolate, (2.5)5–20 × (2)3–11 cm., coriaceous; apex shortly acuminate to acute, sometimes to obtuse; base cordate, occasionally truncate or ± broadly cuneate; margin entire; both surfaces glabrous; lateral veins 8–11 pairs, the (main) basal pairs ± distinctly branched, almost straight and thus not running parallel to the margin, the other veins often furcate far from the margin. Petiole 0.5–4 cm. long, 1–3 mm. thick; stipules 0.5–1 cm. long, densely yellowish-tomentose to subvelutinous or glabrous, caducous. Figs in pairs in the leaf axils or just below the leaves, subsessile or on peduncles up to 0.5 mm. long; basal bracts c. 2 mm. long; receptacle subglobose, 1–2 cm. in diam. when fresh, 0.5–1 cm. in diam. when dry, minutely white to brown puberulous or partly hirtellous, sometimes densely tomentose or pubescent, whitish pink to pale or dark purple at maturity; wall wrinkled when dry.

Botswana. SE: Otse (Ootsi) Hills, c. 40 km. SE. of Gaborone, 30.vi.1974, *Mott* 294 (K; UBLS). **Zambia**. N: Mporokoso Distr., L. Cheshi (Chisi), Mweru-Wantipa Escarpment, c. 1000 m., 13.xii.1960, *Richards* 13685 (K). W: Chingola, 3.x.1955, *Fanshawe* 2486 (K; LISC). C: 1 km. N. of Kafue Bridge, 17.xi.1982, *Berg & Bingham* 1368 (K; U). E: Petauke Distr., Great East Road, c. 130 km. (mile 81) from Chipata on road to Lusaka, 24.v.1952, *White* 2874 (FHO; K). S: Livingstone Distr., Zambezi R. just above Victoria Falls, 27.viii.1947, *Brenan & Greenway* 7765 (FHO; K). **Zimbabwe**. N: Centenary Distr., Muzarabani, Musengezi (Masangedsi) R., near foot of the escarpment, 9.vi.1965, *Bingham* 1446 (PRE; SRGH). W: Bulawayo Distr., Hillside Farm, 12.iii.1954, *Orpen* 16/54 (K; LISC; SRGH). C: Chirumanzu Distr., between Umvuma and Gweru, near Excelsior Mine, 28.ii.1951, *Greenhow* 29/51 (K; SRGH). E: Mutare Distr., Zimunya Reserve, road to Glen Shiel, 31.viii.1958, *Chase* 6984 (K; LISC; SRGH). S: Chibi Distr., c. 14 km. S. of Chibi Admin. Centre, c. 900 m., 6.v.1970, *Biegel & Pope* 3272 (K; SRGH). **Malawi**. N: Nkhata Bay, White Father's Beach, 7.xii.1975, *Pawek* 10394 (K; LMA; NY; PRE; SRGH; UC). C: Dedza Distr., E. side of Chiwao Hill, 14.xii.1966, *Jeke* 39 (K; SRGH). S: Mulanje Mt., 11.vi.1957, *Chapman* 380 (BM; FHO; K; PRE). **Mozambique**. Z: S. of Gúruè, 6.vii.1942, *Hornby* 2693 (PRE). T: Entre Casula e Furancungo, 14.x.1943, *Torre* 6028 (COI; K; LISC; LMA). MS: Manica, R. Mucombre, c. 10 km. from Dombe, 26.x.1953, *Pedro* 4431 (LMA; PRE). M: Goba, prox. rio Maiuana, 3.xi.1960, *Balsinhas* 188 (K).

Widespread in subsaharan Africa from Senegal to Somalia and the southern Arabian Peninsula, and southwards to Namibia and S. Africa (Cape Province). Deciduous woodland and riverine fringes, often on rocks in rocky outcrops, sometimes a strangling fig; 0–2100 m.

This species is rather variable and sometimes difficult to distinguish from the broad-leaved forms of *F. cordata* subsp. *salicifolia*.

10. **Ficus cordata** Thunb., Diss. Fic.: 8, with plate (1786). —Hutch. in F.T.A. **6**, 2: 119 (1916); in F.C. **5**, 2: 530 (1925). —K. Coates Palgrave, Trees Southern Africa: 106, with photograph (1977). —C.C. Berg et al. in Fl. Cameroun **28**: 152, t. 50 (1985). —C.C. Berg in F.T.E.A., Moraceae: 60 (1989); in Kirkia **13**: 259 (1990). Type from S. Africa.

   Urostigma cordatum (Thunb.) Gasp., Ricerche Caprifico: 82 (1845). Type as above.
   Urostigma thunbergii Miq. in Hook., Lond. Journ. Bot. **6**: 556 (1847). Type as for *F. cordata*.

Trees up to 15(35) m. tall. Leafy twigs 2–5 mm. thick, glabrous to pubescent, indumentum white. Leaf lamina lanceolate to ovate or oblong, elliptic or cordiform, rarely obovate, 2–17 × 1–6 cm., coriaceous; apex acuminate, acute, obtuse or occasionally rounded; base cordate to rounded, or obtuse, occasionally cuneate; margin entire; both surfaces glabrous; lateral veins 6–12 pairs, the basal pair unbranched, usually running almost parallel to the margin, other lateral veins furcate near the margin, smaller veins prominent and conspicuous beneath. Petiole 0.5–4(6) cm. long, 1–2 mm. thick, (lamina length: petiole length = 5–8: 1); stipules 0.5–1.5 cm. long, glabrous or ciliolate. Figs mostly in pairs in the leaf axils or just below the leaves, or 2–5 together on spurs 2–3 mm. long on older wood, peduncle 0–3 mm. long; basal bracts c. 1–1.5 mm. long, sometimes caducous. Receptacle subglobose, 5–10 mm. in diam. when fresh, 5–8 mm. in diam. when dry, sparsely and minutely puberulous to glabrous, maturing from green to whitish to dark purple or dark red; wall usually smooth when dry.

Subsp. **cordata** —C.C. Berg in Kew Bull. **43**: 81 (1988); in Kirkia **13**: 259 (1990). —van Greuning in S. Afr. J. Bot. **56**, 6: 607, fig. 5 (1990).

   Ficus tristis Kunth & Bouché, in Ind. Sem. Hort. Berol. **1846**: 19 (1847). Type a cultivated specimen from Berlin Botanical Garden.
   Ficus welwitschii Warb. in Engl., Bot. Jahrb. **20**: 160 (1894). Lectotype from Angola, chosen by Berg in Fl. Cameroun **28**: 152 (1985).
   Ficus welwitschii var. *beröensis* Hiern, Cat. Afr. Pl. Welw. **1**, 4: 999 (1900). Lectotype from Angola, chosen by Berg in Fl. Cameroun **28**: 152 (1985).

Ficus cordata var. *tristis* (Kunth & Bouché) Warb. in Viert. Nat. Ges. Zürich **51**: 137 (1906).
Ficus cordata var. *marlothii* Warb. in Viert. Nat. Ges. Zürich **51**: 137 (1906). Type from Namibia.
Ficus cordata var. *fleckii* Warb. in Viert. Nat. Ges. Zürich **51**: 138 (1906). Type from Namibia.
Ficus rupium Dinter, Deutsch. SW.-Afr. Fl. For. Landt. Frag. 54 (1909). Described from
Namibia, type not designated.

Leafy twigs usually ± densely puberulous to pubescent. Leaf lamina cordiform to ovate
or elliptic, (2)4–9 × (1)2–4.5 cm.; apex acuminate to subcaudate; base cordate to truncate
or rounded (cuneate); petiole 1–2(6) cm. long; stipules 0.2–1 cm. long, up to 5 cm. long on
new flush, usually puberulous to pubescent. Figs usually sessile.

Botswana. N: Aha Hills, 19°41′S, 21°04′E, XaiXai to Quangwa Road., 27.iv.1980, *P.A. Smith* 3462
(K; SRGH).
Also in S. Africa (SW. Cape Province), Namibia and Angola. Dry semi-desert, associated with
dolomite rocks.

Subsp. **salicifolia** (Vahl) C.C. Berg in Kew Bull. **43**: 82 (1988); in F.T.E.A., Moraceae: 63 (1989); in
Kirkia **13**: 260 (1990). —van Greuning in S. Afr. J. Bot. **56**, 6: 607, fig. 6 (1990). TAB. **19**. Type from
Yemen.
Ficus indica sensu Forssk., Fl. Aegypt.-Arab.: 179 (1775) non L. (1753).
Ficus religiosa sensu Forssk., Fl. Aegypt.-Arab.: 170 (1775) non L. (1753).
Ficus salicifolia Vahl, in Symb. Bot. **1**: 82, t. 23 (1790). —Hutch. in F.T.A. **6**, 2: 115 (1916). —K.
Coates Palgrave, Trees Southern Africa: 112, with photograph (1977). Type from Yemen.
Urostigma salicifolium (Vahl) Miq. in Hook., Lond. Journ. Bot. **6**: 556 (1847). Type from
Yemen.
Ficus salicifolia var. *australis* Warb. in Viert. Nat. Ges. Zürich **51**: 139 (1906). Type from S.
Africa.
Ficus pretoriae Burtt Davy in Trans. Roy. Soc. S. Afr. **2**: 365 (1912). —Hutch. in F.T.A. **6**, 2: 116
(1916); in F.C. **5**, 2: 528 (1925). —F. White, F.F.N.R.: 32 (1962). Types from S. Africa.

Leafy twigs usually minutely puberulous. Leaf lamina subovate to oblong or lanceolate,
sometimes ovate or elliptic, 2–17 × 1–5.5 cm.; apex acute to ± acuminate (rounded); base
rounded to cordate or obtuse (cuneate); petiole 0.5–4(6.5) cm. long; stipules 0.5–1.5 cm.
long, glabrous or ciliolate. Figs on peduncles up to 3 mm. long, sometimes subsessile.

Botswana. SE: Kanye, fr. 6.vii.1980, *Woollard* 761 (SRGH). **Zambia**. C: Mumbwa, 31.v.1961,
Fanshawe 6634 (K). E: Petauke, Great East Road, c. 130 km. (Mile 81) from Chipata (Fort Jameson) to
Lusaka, 24.v.1952, *White* 2875 (FHO; K; NDO). **Zimbabwe**. N: Makonde Distr., Whindale Farm, near
Mhangura (Mangula), 1.ix.1963, *Jacobsen* 2194 (PRE). W: Matobo Distr., Farm Besna Kobila, iii.1957,
Miller 4209 (K; SRGH). C: Harare Distr., Norton, by Munyame (Hunyani) R., 4.i.1948, *Wild* 2271
(K). E: Mutare, West base Murahwa's Hill, 17.ii.1969, *Chase* 8525 (LISC; PRE; SRGH). S: Beitbridge
Distr., Shashi-Limpopo confluence, 12.v.1959, *Drummond* 6120 (COI; K; SRGH). **Malawi**. S:
Mangochi, 17.ix.1929, *Burtt Davy* 21755 (FHO). **Mozambique**. T: Mágoè para Chicoa, morro à
esquerda, 25.ii.1970, *Torre & Correia* 18093 (K; LISC; LMA; LMU). M: Junto a ponte de Goba,
16.xi.1940, *Torre* 2060 (LISC).
Also known from Chad and Algeria, Egypt and W. Saudi Arabia, NW. Uganda and NE. Zaire and
extends to S. Africa (Natal and Transvaal). Woodland, often amongst rocks and on rocky outcrops,
ironstone, limestone, sandstone and granite; 2700 m.
Subsp. **lecardii** (Warb.) C.C. Berg occurs in W. Africa from Senegal through N. Cameroons to the
Central African Republic. It differs from subsp. *cordata* in having figs mostly pedunculate and twigs
often glabrous, and from subsp. *salicifolia* in having broader leaves and petioles which usually dry
dark brown to blackish.

11. **Ficus verruculosa** Warb. in Engl., Bot. Jahrb. **20**: 166 (1894). —Hutch. in F.T.A. **6**, 2: 114 (1916).
—Lebrun & Boutique in F.C.B. **1**: 120 (1948). —Keay in F.W.T.A. ed. 2, **1**: 607 (1958). —F. White,
F.F.N.R.: 32, t. 6E (1962). —K. Coates Palgrave, Trees Southern Africa: 118 (1977). —C.C. Berg et
al. in Fl. Cameroun **28**: 154, t. 51 (1985). —C.C. Berg in F.T.E.A., Moraceae: 63 (1989); in Kirkia
13: 260 (1990). —van Greuning in S. Afr. J. Bot. **56**, 6: 607, fig. 8 (1990). Type from Angola.
Ficus praeruptorum Hiern, Cat. Afr. Pl. Welw. **1**, 4: 1004 (1900). Type from Angola.

Shrub or weak-stemmed, sparsely branched small tree, mostly to 2.5 m. tall, sometimes
up to 5 m., occasionally up to 12 m. tall. Leafy twigs 1–5 mm. thick, glabrous or densely
white hirtellous to subtomentellous. Leaf lamina oblong to lanceolate, 3.5-10(20) ×
1.5-3.5(8.5) cm., coriaceous; apex subacute to obtuse; base obtuse or rounded to
subcordate; margin entire; both surfaces glabrous; lateral veins (8)10–16 pairs, the basal
pair unbranched, running almost parallel to the margin, the other veins furcate near the
margin, midrib prominent beneath, the other veins plane and inconspicuous; petiole
5-20(30) mm. long, 1-2 mm. thick, (lamina length: petiole length = (5)8–10: 1); stipules

Tab. 19. FICUS CORDATA subsp. SALICIFOLIA. 1, fertile twig (×⅔), *Chase* 2722; 2, fertile twig (×⅔), *Moss* 2142; 3, fertile twig with sessile figs (×⅔), *Dinter* 275; 4, 1/s through fig (× 3), *Hutchinson & Gillett* 4422. Drawn by M. Tebbs.

0.5–3.5(4) cm. long, glabrous or densely white puberulous, caducous. Figs in pairs, sometimes 4 together, in the leaf axils or just below the leaves; peduncle 2–5(10) mm. long, 1–1.5 mm. thick; basal bracts c. 1 mm. long. Receptacle subglobose to ellipsoid, 0.5–2 cm. in diam. when fresh, 0.5–1(1.2) cm. in diam. when dry, glabrous or minutely puberulous, dark purple or dark red at maturity; wall often wrinkled when dry.

Botswana. N: Okavango Swamps, Moremi Wildlife Reserve, island on Gobega Lagoon, 5.iii.1972, *Biegel & Russell* 3845 (K; LISC; SRGH). **Zambia**. B: 16 km. N. of Senanga, 3.vii.1952, *Codd* 7295 (BM; K; PRE). N: Between Serenje & Mpika, 16.vii.1930, *Pole-Evans* 2918(25) (K; PRE). W: Mwinilunga Distr., by Matonchi R., below Dam, 21.x.1937, *Milne-Redhead* 2889 (K; PRE). C: Kabwe (Broken Hill) to Chiwefwe, 14.vii.1930, *Hutchinson & Gillett* 3649 (BM; K; LISC; PRE). E: Forest Reserve W. of Nyimba, 12.xii.1958, *Robson* 928 (K; LISC). S: Mazabuka, near Siamambo Stream, Siamambo Forest Reserve, Choma, 23.vii.1952, *Angus* 11 (FHO; K; M?). **Zimbabwe**. N: Guruve Distr., Nyamunyeche Estate, 28.ix.1978, *Nyariri* 380 (K; SRGH). W: Hwange, Special Native Area A, point where Mondi R. drops into Zambezi Gorge, viii.1956, *Davies* 2060 (K; PRE; SRGH). C: Harare Distr., Twentydales, 9.ix.1946, *Wild* 1214 (K; SRGH). E: Mutare Distr., Tsonzo Division, Kukwanisa, 8.xii.1967, *Biegel* 2361 (K; LISC; PRE; SRGH). S: c. 4 km. E. of Great Zimbabwe, 1.vii.1930, *Hutchinson & Gillett* 3391 (BM; K). **Malawi**. N: Mzimba Distr., towards Lunyangwa from Marymount, Mzuzu, 11.xii.1970, *Pawek* 4084 (K; MAL). C: c. 6 km. S. of Dedza, c. 1700 m., 2.ii.1959, *Robson* 1419 (BM; K; LISC; PRE). S: Shire Highlands, near Kankanje, 26.ix.1859, *Kirk* s.n. **Mozambique**. N: Maniamba, andados 45 km. de Lichinga (Vila Cabral) para Unango, 2.iii.1946, *Torre & Paiva* 10950 (LISC; LMA; UC). Z: Gúruè, serra do Gúruè, cascata do rio Licungo, a. 20 km. da fabrica Mocambique em direcão ao regulo Mgunha, 9.xi.1967, *Torre & Correia* 16016 (LISC). T: Macanga Distr., prox. de Furancungo, 29.ix.1942, *Mendonça* 514 (LISC). MS: Zona R., 1906, *Swynnerton* 146a (BM). GI: Inhambane, 1908, *Sim* 5414 (K). M: Bilene Distr., prox. de Macia, 29.iii.1948, *Torre* 7579 (EA; LISC; LMA).

Widespread from Nigeria to Uganda, Kenya and Tanzania, and southwards to Angola, Namibia and S. Africa (Natal). Riverine vegetation, grassland near water and swamp forest margins, sometimes forming thickets.

12. **Ficus bussei** Mildbr. & Burret in Engl., Bot. Jahrb. **46**: 213 (1911). —Friis in Nordic Journ. Bot. **5**: 332 (1985). C.C. Berg in FTEA., Moraceae: 64 (1989); in Kirkia **13**: 262 (1990). Type from Tanzania.
    Ficus fasciculata Warb. in Engl., Bot. Jahrb. **20**: 175 (1894) non Benth. (1873), nom. illegit. Type from Tanzania.
    Ficus changuensis Mildbr. & Burret in Engl., Bot. Jahrb. **46**: 212 (1911). Type as for *F. fasciculata* Warb.
    Ficus zambesiaca Hutch. in Bull. Misc. Inf., Kew **1915**: 341 (1915); in F.T.A. **6**, 2: 198 (1917). —F. White, F.F.N.R.: 33 (1962). —K. Coates Palgrave, Trees Southern Africa: 119 (1977). Type: Malawi, Shire Valley, Katunga, *Scott* s.n. (K, lectotype, chosen by C.C. Berg in F.T.E.A., Moraceae: 64 (1989).

Tree up to 20 m. high, epiphytic and secondarily terrestrial. Leafy twigs 4–12 mm. thick, sparsely sometimes densely puberulous to hirtellous or glabrous, periderm hardly flaking off when dry. Leaves spirally arranged; lamina subovate to oblong, 5–24 × 3–9.5(11.5) cm., coriaceous, often brittle when dry; apex subacute to obtuse; base cordate; margin entire to repand; superior surface glabrous, or puberulon on the midrib, inferior surface ± glabrous or sparsely hirtellous to puberulous, midrib subhirsute; lateral veins (8)10–16 pairs, the basal pair reaching the margin far below the middle of the lamina, tertiary venation partly scalariform; petiole 2–8 cm. long, 2–4 mm. thick, epidermis not flaking off; stipules 3–12 mm. long, up to 30 mm. long on new flush, glabrous or pubescent at the base, caducous. Figs in pairs or solitary in the leaf axils; peduncle 10–25 mm. long, recurved; basal bracts c. 3 mm. long, persistent. Receptacle subglobose to ellipsoid, 2–3 cm. in diam. when fresh, 1–1.5 cm. in diam. when dry, puberulous, smooth or verruculate, greenish at maturity.

Zambia. S: Siavonga-Lusaka road, near Mbendele R., 17.xi.1982, *Berg & Bingham* 1376 (K; U). **Zimbabwe**. N: Kariba Gorge on the Zambezi R. bank, ix.1960, *Goldsmith* 107/60 (K; PRE; SRGH). **Malawi**. N: Mzimba Distr., 16 km. W. of Mzambazi, 27.xi.1969, *Pawek* 3030 (K). C: Ntchisi Island, Chirwa, 13.vii.1966, *Agnew et al.* s.n. (MAL). S: Nsanje Distr., near Tengani Health Centre turnoff, 19.v.1983, *Banda & Balaka* 1976 (K; MAL). **Mozambique**. N: Niassa Prov., Marrupa, Experimental Field Mademo, 18.ii.1982, *Jansen* 7810 (K). Z: 50 km. NE. of Mopeia Velha on road to Quelimane, 7.xii.1971, *Müller & Pope* 1944 (K; LISC; SRGH). T: Prox. de Chicoa (Chioco), R. Luia, 26.ix.1942, *Mendonça* 443 (K; LISC; LMU; PRE). MS: Changara, posto de Fronteira, 13.x.1954, *Barbosa* 5595 (BM; LMJ). M: Matutuine (Bela Vista), entre Zitundo a Ponta do Ouro, 8.iii.1968, *Balsinhas & Gomes e Sousa* 5065 (PRE).

Extending down the E. coast of Africa from Somalia to Mozambique and inland up the Zambezi Valley. Coastal and low altitude riverine and woodland vegetation, often on alluvial soils, usually as a strangler fig; 0–600 m.

13. **Ficus wakefieldii** Hutch. in Bull. Misc. Inf., Kew **1915**: 335, with plate (1915); in F.T.A. **6**, 2: 168 (1916). —F. White, F.F.N.R.: 32 (1962). —C.C. Berg in F.T.E.A., Moraceae: 65 (1989); in Kirkia **13**: 263 (1990). Lectotype from Kenya, chosen by Berg in F.T.E.A., Moraceae: 65 (1989).

Tree up to 25 m. tall, epiphytic, secondarily terrestrial, with a wide crown. Leafy twigs (3)5–12 mm. thick, with minute to short hairs, intermixed with much longer yellow to brownish hairs, periderm flaking off when dry. Leaves spirally arranged; lamina cordate to ovate or broadly elliptic to subcircular or subreniform, sometimes broadly obovate, 6–23 × 5–23 cm., subcoriaceous; apex rounded, sometimes very shortly and obtusely acuminate; base cordate; margin subentire; superior surface sparsely hirtellous to subhirsute, inferior surface sparsely to densely hirtellous to puberulous, the main veins yellow hirsute; lateral veins 5–8 pairs, the main basal pair branched and reaching the margin at or just below the middle of the lamina, tertiary venation partly scalariform; petiole 2–5.5(9) cm. long, (1.5)2–4(5) mm. thick, epidermis flaking off when dry; stipules 0.5–1.5 cm. long, up to 4 cm. long on new flush, yellow to brownish hirsute or subsericeous, caducous. Figs in pairs in the leaf axils, sessile, initially enclosed within ovoid, calyptrate buds, up to 15 mm. long; basal bracts 3–5 mm. long, persistent. Receptacle subglobose, (1.2)1.5–2 cm. in diam. when fresh, (0.8)1–1.5 cm. in diam. when dry, densely white to yellow pubescent, subhirsute or sparsely hirtellous.

Zambia. N: Lake Mweru, 13.xi.1957, *Fanshawe* 3944 (K). W: Ndola Golf Course, 23.x.1952, *Angus* 654 (BM; FHO; K; PRE). C: Kabwe, 22.xi.1982, *Berg & Bingham* 1388 (K; U).

Also known from Zaire (Shaba), Tanzania, Kenya and Uganda. Riverine forest and lake shores and miombo woodland, often on termite mounds and in rocky places; 700–2000 m.

14. **Ficus glumosa** Del., Cent. Pl. Méroé: 63 (1826). —Hutch. in F.T.A. **6**, 2: 171 (1916). —Lebrun & Boutique in F.C.B. **1**: 147 (1948). —Keay in F.W.T.A. ed. 2, **1**: 609 (1958). —K. Coates Palgrave, Trees Southern Africa: 108 (1977). —Aweke in Meded. Landb. Wag. **79**-3: 25, fig. 6 (1979). —C.C. Berg et al. in Fl. Cameroun **28**: 170, t. 58 (1985). —C.C. Berg in F.T.E.A., Moraceae: 65 (1989); in Kirkia **13**: 263 (1990). —van Greuning in S. Afr. J. Bot. **56**, 6: 612, fig. 11 (1990). Type from Ethiopia.

Urostigma glumosum (Del.) Miq. in Hook., Lond. Journ. Bot. **6**: 552 (1847). Type as above.

Sycomorus hirsuta Sond. in Linnaea **23**. 137 (1850) non *Ficus hirsuta* Vell. (1825). Type from S. Africa.

Ficus sonderi Miq. in Ann. Mus. Bot. Lugd.-Bat. **3**: 295 (1867). —Hutch. in F.T.A. **6**, 2: 170 (1916); in F.C. **5**, 2: 534 (1925). —Lebrun & Boutique in F.C.B. **1**: 144 (1948). —F. White, F.F.N.R.: 32 (1962). —K. Coates Palgrave, Trees Southern Africa: 114, with photograph (1977). Type as for *Sycomorus hirsuta*.

Ficus barbata Warb. in Engl., Bot. Jahrb. **20**: 168 (1894) non Wallich (1831). Type from Angola.

Ficus rukwaensis Warb. in Engl., Bot. Jahrb. **30**: 295 (1901). Type from Tanzania.

Ficus rehmannii Warb. in Viert. Nat. Ges. Zürich **51**: 136 (1906). Type from S. Africa.

Ficus rehmannii var. *ovatifolia* Warb. in Viert. Nat. Ges. Zürich **51**: 136 (1906). Type from S. Africa.

Ficus rehmannii var. *villosa* Warb. in Viert. Nat. Ges. Zürich **51**: 137 (1906). Type: Zimbabwe, Matopos, *Marloth* 3382 (B, holotype).

Ficus montana Sim, For. Fl. Port. E. Afr.: 101, t. 95A (1909). Type: Mozambique, Lebombo Mts., *Sim* 6313 (K, holotype).

Ficus gombariensis De Wild. in Fedde, Repert. **12**: 199 (1913). Type from Zaire.

Ficus kitaba De Wild. in Bull. Soc. Roy. Bot. Belge **52**: 215 (1914). Type from Zaire.

Ficus glumosoides Hutch. in Bull. Misc. Inf., Kew **1915**: 336 (1915). Type from Angola.

Trees up to 10 m. tall or shrubs, terrestrial, branches spreading. Leafy twigs 2–6 mm. thick, indumentum dense, of short white hairs intermixed with much longer yellow to whitish hairs especially on the nodes, or glabrous; periderm of older parts flaking off when dry. Leaves spirally arranged; lamina oblong to broadly elliptic or broadly ovate, sometimes obovate or subcircular, 2–14(19) × 1.2–9.5(13) cm., subcoriaceous; apex shortly acuminate to subacute or subobtuse; base cordate, sometimes rounded; margin entire; superior surface puberulous, hirtellous or subtomentose, sometimes almost glabrous, inferior surface subtomentose, main veins yellow hirsute or glabrous; lateral veins 3–7 pairs, the basal pair faintly branched, usually ending at the margin far below the middle of the lamina, or sometimes at or above the lamina middle, tertiary venation reticulate or tending to scalariform, the smaller veins (reticulum) inconspicuous beneath; petiole 0.5–4(8) cm. long, 1–2(2.5) mm. thick, epidermis not flaking off; stipules 5–15 mm. long, up to 40 mm. long on new flush, caducous, partly or ± sparsely hirsute to subsericeous, the indumentum bright yellow or sometimes whitish. Figs in pairs in the leaf axils or 0–10 mm.

below the leaves, subsessile or sometimes on peduncles up to 3 mm. long; basal bracts c. 3 mm. long, persistent. Receptacle globose to ellipsoid, c. 1–1.5 cm. in diam. when fresh, 0.5–1 cm. in diam. when dry, densely tomentose to pubescent or almost glabrous, orange to red or pink at maturity, often with darker spots.

Zambia. N: Mbala Distr., Lake Tanganyika, mouth of Kalambo R., 29.ix.1963, *Richards* 18227 (K). W: Mwinilunga Distr., E. of Matonchi Farm, 29.x.1937, *Milne-Redhead* 3000 (BM; K; PRE). C: Luangwa Distr., Rufunsa-Shikabeta road., 1.i.1978, *Bingham* 2620 (K). E: Chipata, 2.vi.1958, *Fanshawe* 4505 (K). S: Kafue River Gorge, 6.x.1957, *Angus* 1727 (FHO; K). **Zimbabwe**. N: Makonde Distr., Chirombodzi Farm, Mhangura (Mangula) Area, 7.x.1962, *Jacobsen* 1824 (PRE). W: Matopos, viii.1930, *Hutchinson* 4137 (K). C: Chirumanzu (Chilimanzi) T.T.L., 7.iii.1951, *Greenhow* 39/51 (K; SRGH). S: Masvingo (Fort Victoria), Tokwe R., 7.x.1949, *Wild* 3023 (K). **Malawi**. N: Mzimba Distr., Hora Mt., 19.i.1978, *Pawek* 13624 (K; MAL; MO; SRGH; UC). C: Kasungu Distr., Chipala Hill, Chamama, c. 1050 m., 16.i.1959, *Robson & Jackson* 1213 (K). S: Mulanje Distr., Nachilonga Hill, S. of Likanya Tea Estate Bay, 25.iv.1987, *J.D. & E.G. Chapman* 8453 (K; MO). **Mozambique**. N: Ribáuè, Posto Agrícola, 5.xi.1942, *Mendonça* 1266 (K; LISC; PRE). M: Estrada para Goba, a 10 km. depois do cruz de estradas Goba-Namaacha-Maputo, 3.x.1961, *Balsinhas* 515 (COI; K; PRE).

Widespread in subsaharan Africa, extending into W. Saudi Arabia in the east and into S. Africa (Natal) and Namibia in the south. Rock outcrops and rocky slopes, less often in riverine and open miombo woodland occasionally on termite mounds.

A very variable species, especially in regard to its indumentum. The indumentum is usually densest on trees in the southern part of its range of distribution. Almost glabrous specimens are more common in NE. Africa, but are also recorded for Angola and Namibia.

F. glumosa may be confused with *F. stuhlmannii* but can be distinguished by the reticulum which is plane on the leaf lower surface and by the presence of some yellowish hairs on the twigs.

15. **Ficus stuhlmannii** Warb. in Engl., Bot. Jahrb. **20**: 161 (1894). —Hutch. in F.T.A. **6**, 2: 170 (1916); in F.C. **5**, 2: 536 (1925). —F. White, F.F.N.R.: 33 (1962). —K. Coates Palgrave, Trees Southern Africa: 115 (1977). —C.C. Berg in F.T.E.A., Moraceae: 66 (1989); in Kirkia **13**: 263 (1990). —van Greuning in S. Afr. J. Bot. **56**, 6: 614, fig. 12 (1990). Type from Tanzania.

Ficus howardiana Sim, For. Fl. Port. E. Afr.: 100, t. 92A (1909). Type: Mozambique, Mubalusi, *Sim* 6262 (K, holotype).

Ficus homblei De Wild. in Fedde, Repert. **12**: 195 (1913). Type from Zaire.

Ficus dar-es-salaamii Hutch. in F.T.A. **6**, 2: 171 (1916). Type from Tanzania.

Tree up to 10(15) m. tall, terrestrial or hemi-epiphytic and strangling. Leafy twigs (2)4–8 mm. thick, ± densely white puberulous to hirtellous, pale yellow hirsute on the nodes, periderm of older parts ± flaking off when dry. Leaves spirally arranged; lamina oblong to elliptic or ± ovate to subobovate, sometimes subcircular, 2.5–18 × 1–8 cm., subcoriaceous; apex rounded to subacute, sometimes shortly and obtusely acuminate; base cordate to rounded; margin subentire; superior surface puberulous to hirtellous, inferior surface densely hirtellous to subtomentose on the veins; lateral veins (3)4–7 pairs, the basal pair of veins not or faintly branched, usually ending at the margin well below the middle of the lamina, tertiary venation reticulate and prominent; petiole 0.5–4 cm. long, (1)2–3 mm. thick, epidermis not flaking off; stipules 0.5–1.5 cm. long, white to pale yellow subsericeous to subhirsute or puberulous, caducous. Figs in pairs in the leaf axils, sometimes also just below the leaves, subsessile; basal bracts c. 3 mm. long. Receptacle globose to ellipsoid, 15–22 mm. in diam. when fresh, (7)10–18 mm. in diam. when dry, densely white pubescent to sparsely puberulous, pinkish or purplish at maturity.

Zambia. N: Mbala Distr., Mpulungu, N'mkole, 20.ii.1964, *Richards* 19048 (K). W: Ndola township near Government Rest House, 24.x.1952, *Angus* 655 (BM; FHO; K). C: Mt. Makulu Research Station, 16 km. S. of Lusaka, 11.xi.1956, *Simwanda* 72 (K; SRGH). S: Mazabuka Distr., Tsiknaki's Farm, 21 km. N. of Choma, 30.i.1960, *White* 6648 (FHO; K). **Zimbabwe**. N: Shamva Distr., Chipoli Farm, 1.v.1959, *Moubray* 79 (K; SRGH). C: Nyamatendera R., vii.1970, *Burrows* 448 (SRGH). E: Nyanga Distr., Nyamaropa T.T.L. Regina Coeli Mission Orchard, 16.i.1967, *Biegel* 1766 (K; SRGH). **Malawi**. N: Mzimba Distr., 8 km. SE. of Mzambazi, 1.ii.1976, *Pawek* 10819 (K; LMA; MAL; PRE; SRGH; UC). **Mozambique**. N: Cabo Delgado, Pemba, entre Metuge e Mahate, 3.x.1948, *Barbosa* 2338 (LISC). Z: Entre Morrumbala e Megaza a 37.2 km. de Morrumbala, 13.vi.1949, *Barbosa & Carvalho* 3071 (K; M). T: Mutarara Valley, 20.x.1971, *Hafferri* 17 (SRGH). MS: Gorongosa, Parque Nacional de Caja, no Acampamento de Chitengo, 13.xi.1963, *Torre & Paiva* 9225 (LISC; LMA; LMU; UC). M: Maputo Distr., entre Magude & Panjane, 27.i.1948, *Torre* 7226 (EA; K; LISC; LMU; SRGH).

Also in Uganda, Kenya, Tanzania, Zaire and S. Africa. Miombo and mixed deciduous woodland, in low altitudes, often on termitaria or on rocky outcrops; 0–1800 m.

16. **Ficus nigropunctata** Mildbr. & Burret in Engl., Bot. Jahrb. **46**: 220, t. 3 (1911). —Hutch. in F.T.A. **6**, 2: 173 (1916). —K. Coates Palgrave, Trees Southern Africa: 111 (1977). —C.C. Berg in F.T.E.A., Moraceae: 67 (1989); in Kirkia **13**: 263 (1990). Lectotype from Tanzania, chosen by Berg in F.T.E.A., Moraceae: 67 (1989).

Shrub or tree up to 7 m. tall, terrestrial or sometimes hemi-epiphytic and strangling. Leafy twigs 1–3(6) mm. thick, puberulous to hirtellous or subtomentellous, periderm not flaking off, bark of the older wood often blackish and conspicuously lenticellate. Leaves spirally arranged; lamina oblong to elliptic or ± obovate, sometimes ovate, 1–9.5 × 0.6–5.5 cm., chartaceous to subcoriaceous; apex shortly acuminate to subacute; base rounded to cordate; margin crenulate; superior surface puberulous to hirtellous or hispidulous, sometimes black punctate when dry, inferior surface puberulous to hirtellous on the veins; lateral veins 3–5(6) pairs, the basal pairs faintly branched, usually ending near the margin well below the middle of the lamina, tertiary venation reticulate; petiole 3–20 mm. long, 0.5–1 mm. thick, epidermis not flaking off; stipules 2–12 mm. long, sparsely puberulous to hirtellous, caducous. Figs in pairs in the leaf axils, also on older wood, sessile; basal bracts 2–2.5 mm. long, persistent. Receptacle subglobose, 1–1.2 cm. in diam. when fresh, 0.5–1 cm. in diam. when dry, puberulous to hirtellous, green with red spots or reddish at maturity.

Botswana. N: Linyanti, 1100 m., *Miller* B/1031 (SRGH). **Zambia**. C: c. 35 km. from Lusaka to Kabwe, 19.xi.1982, *Berg & Bingham* 1382 (K; U). E: Great East Road, Forest Reserve, W. of Nyimba, 12.xii.1958, *Robson* 928 (BM; K; SRGH). S: Mazabuka Distr., c. 30 km. N. of Pemba near Milimo Village, 12.xi.1960, *White* 6967 (FHO; K; SRGH). **Zimbabwe**. N: Gokwe Distr., Sengwa Res. Station, 25.xi.1976, *P. Guy* 2456 (K; SRGH). W: Hwange Distr., Matetsi Safari Area Headquaters, 10.iv.1979, *Gonde* 232 (K; SRGH). C: Chegutu, 31.v.1951, *Hornby* 3264 (SRGH). **Malawi**. N: Karonga Distr., Stevenson's Road., c. 35 km. W. of Karonga, 26.iv.1977, *Pawek* 12702 (K; PRE; SRGH). **Mozambique**. N: Lalaua, andados 10 km. de para Ribáuè, 22.i.1964, *Torre & Paiva* 10114 (LISC). Also known from Kenya, Tanzania and S. Africa (Natal). Dry mixed deciduous woodland, usually on rocky outcrops; 0–1300 m.

Similar to *F. stuhlmannii* but *F. nigropunctata* may be distinguished by its smaller, thinner leaves, by its smaller figs, and by the dried older wood which is blackish with conspicuous lenticels.

17. **Ficus tettensis** Hutch. in Bull. Misc. Inf., Kew **1915**: 341 (1915); in F.T.A. **6**, 2: 199 (1917). —K. Coates Palgrave, Trees Southern Africa: 116 (1977). —C.C. Berg in Kirkia **13**: 264 (1990). —van Greuning in S. Afr. J. Bot. **56**, 6: 614, fig. 13 (1990). Type: Mozambique, Tete (Tette), *Kirk* s.n. (K, holotype).
Ficus smutsii Verdoorn in Bull. Misc. Inf., Kew **1935**: 205 (1935). Type from S. Africa.

Tree up to 6 m. tall, or shrub, mostly epilithic; bark of the trunk orange-yellow. Leafy twigs 2–6 mm. thick, white hirtellous to pubescent, periderm of older parts flaking off when dry. Leaves spirally arranged; lamina subreniform to cordate, 1.5–11.5 × 2–12 cm., subcoriaceous; apex short acuminate or subacute to rounded; base cordate; margin subentire or irregularly crenate; superior surface ± densely hirtellous to pubescent, often somewhat scabrous, inferior surface hirtellous, subtomentose on the veins; venation prominent beneath, lateral veins 3–5(6) pairs, the basal pair branched, ending at the margin above the middle of the lamina; tertiary venation reticulate or partly scalariform; petiole 0.5–3(5.5) cm. long, 1–2 mm. thick, epidermis not flaking off; stipules 2–6 mm. long, almost as wide as long, apex rounded, white to brownish pubescent, caducous. Figs in the leaf axils, subsessile or on peduncles up to 2 mm. long, basal bracts 2–3 mm. long, caducous. Receptacle subglobose, 0.5–1 cm. in diam. when dry, densely white puberulous to hirtellous.

Botswana. N: Ngwato Distr., c. 5 km. (3 m.) S. of Topsi siding, xii.1948, *Miller* B/804 (K). SE: Sephare in Mahalapye, 11.iv.1959, *De Beer* 889 (K; SRGH). **Zambia**. S: Mazabuka Distr., Ghoma township, 22.vii.1952, *Angus* 1 (FHO; PRE). **Zimbabwe**. N: Mazowe Distr., Chipoli Farm, 24.ix.1958, *Moubray* s.n. (PRE). W: Matopos Reserve, 4.ii.1954, *Orpen* 8/54 (K; LISC; SRGH). C: Wedza Distr., c. 1.6 km. S. of Ruzawi R., on Inoro road, 22.v.1968, *Rushworth* 1125 (K; SRGH). S: Gwanda Distr., Doddieburn Ranch, junction of Umzingwane & Sibizini (Tsibizini) Rivers, 5.v.1972, *Pope* 635 (K; LISC; PRE; SRGH). **Malawi**. N: Mzimba Distr., 8 km. S. of Euthini, 27.xi.1976, *Pawek* 11977 (LMA; PRE). **Mozambique**. N: Mandimba, Namwera, 16.xii.1941, *Hornby* 2434 (PRE). T: andados c. de 10 km. de Magoé para Chicoa, 25.ii.1970, *Torre & Correia* 18097 (COI; LISC; LMU). MS: Cheringoma, Inhaminga, Ginge, 21.v.1948, *Mendonça* 4343 (LISC; PRE). M: Machava, 6.xii.1963, *Balsinhas* 682 (MAL; PRE).
Also in S. Africa (N. Transvaal), probably in Socotra. Occasionally at lower and medium altitudes on rocky outcrops and in clefts in rocks; 0–1000 m.

18. **Ficus muelleriana** C.C. Berg in Kew Bull. **43**: 85 (1988); in Kirkia **13**: 264 (1990). Type: Mozambique, Manica e Sofala, 1 km. E. of Makurupini Falls, *Rich* s.n. (SRGH, holotype).

Shrub. Leafy twigs 2.5–3 mm. thick, densely white puberulous, when dry red-brown, periderm not flaking off. Leaves spirally arranged; lamina ovate, 4.5–9.5 × 3–5 cm., subcoriaceous; apex acute to ± faintly acuminate; base cordate; margin entire; superior surface sparsely puberulous, the midrib more densely so, inferior surface puberulous on the midrib, hirtellous on the lateral veins, tomentose on the smaller veins; venation above obscure, beneath prominent, lateral veins 5–8 pairs, the basal pair branched, ending at the margin just below or at the middle of the lamina, tertiary venation reticulate; petiole (1.2)3.5–5.5 cm. long, epidermis not flaking off; stipules 1–1.5 cm. long, minutely puberulous, caducous. Figs in pairs in the leaf axils, subsessile; basal bracts 2, 1.5–2 mm. long, persistent. Receptacle subglobose, c. 5 mm. in diam. when dry, densely white-puberulous.

Mozambique. MS: Manica Distr., Maronga, 13.viii.1945, *Simão* 462 (LISC).
Not known from elsewhere.
Differs from *F. tettensis* in leaf shape (distinctly longer than broad), in the sparser indumentum on the leaf superior surface, the longer stipules and the persistent periderm of the branches.

19. **Ficus abutilifolia** (Miq.) Miq. in Ann. Mus. Bot. Lugd.-Bat. **3**: 288 (1867). —Hutch. in F.T.A. **6**, 2: 191 (1916). —Keay in F.W.T.A. ed. 2, **1**: 609 (1958). —C.C. Berg et al. in Fl. Cameroun **28**: 162, t. 54 (1985). —C.C. Berg in F.T.E.A., Moraceae: 67 (1989); in Kirkia **13**: 264 (1990). —van Greuning in S. Afr. J. Bot. **56**, 6: 610, fig. 10 (1990). Type from Ethiopia, lectotype chosen by Berg in Fl. Cameroun **28**: 164 (1958).
Urostigma abutilifolium Miq. in Hook., Lond. Journ. Bot. **6**: 551 (1847); in Verh. Eerste Kl. Kon. Ned. Inst. Wet. Amsterdam, ser. 3, **1**: 133, t. 3 (1849). Types from Ethiopia and S. Africa.
Ficus soldanella Warb. in Viert. Nat. Ges. Zürich **51**: 136 (1906). —Hutch. in F.T.A. **6**, 2: 176 (1916); in F.C. **5**, 2: 533 (1925). —K. Coates Palgrave, Trees Southern Africa: 114 (1977). Type from S. Africa.
Ficus picta Sim, For. Fl. Port. E. Afr.: 99, t. 94B (1909). Type: Mozambique, Miluane, *Sim* 6302 (K, holotype).

Tree up to 15 m. tall, terrestrial, often (hemi-)epilithic. Leafy twigs 6–15 mm. thick, glabrous or yellowish to whitish-tomentose or puberulous, periderm often flaking off when dry. Leaves spirally arranged; lamina cordate to broadly ovate or subreniform, 6–19 × 5–20 cm., subcoriaceous; apex shortly acuminate to subacute or obtuse to rounded; base cordate; margin entire; superior surface glabrous or with sparse hairs on the main veins, inferior surface puberulous to subtomentellous, sometimes indumentum only in the axils of the lateral veins, occasionally glabrous; lateral veins 7–9 pairs, the basal pair branched, ending at the margin at or above the middle of the lamina, tertiary venation partly scalariform; petiole 2–10(18) cm. long, 2–4 mm. thick, epidermis not flaking off; stipules 5–20 mm. long, puberulous or glabrous, caducous. Figs in pairs, or sometimes up to 4 together, in the leaf axils or just below the leaves, pedunculate or subsessile; peduncle 3–15 mm. long; basal bracts (2)2.5–3.5 mm. long, persistent, free parts occasionally caducous, leaving a collar-like rim. Receptacle subglobose, obovoid or ellipsoid, 12–20 mm. in diam. when fresh, 5–15 mm. in diam. when dry, sparsely, minutely puberulous, reddish or yellow at maturity.

Botswana. SE: Gaborone, 28.viii.1974, *Mott* 337 (K; SRGH; UBLS). **Zambia**. C: Luangwa Distr., Luangwa (Feira), 5.xii.1968, *Fanshawe* 10460 (K; NDO). E: Lutembwe R. Gorge, East of Machinje Hills, 13.x.1958, *Robson* 97 (BM; K; PRE). S: Siavonga-Lusaka road, near Njami Hill, 17.xi.1982, *Berg & Bingham* 1377 (K). **Zimbabwe**. N: Mazowe Distr., Chiweshe T.T.L., 27.xi.1969, *Orpen* in GHS 199255 (K; PRE; SRGH). W: Hwange Distr., Deka R., 26.xii.1952, *Lovemore* 349 (K; LISC; SRGH). C: Charter Distr., Manesi Range, 15.i.1962, *Wild* 5584 (K; PRE; SRGH). E: Chipinge Distr., Chisumbanje Area, Remayi, 1.ii.1975, *Pope, Biegel & Russell* 1509 (K; PRE; SRGH). S: Beitbridge Distr., Shashi R., c. 11 km. downstream from Tuli Police Camp, 3.v.1959, *Drummond* 6076 (K; PRE; SRGH). **Malawi**. C: Salima Distr., Lake Malawi, Namalenje Island in Senga Bay, 2 km. from Grand Beach Hotel, 21.iii.1977, *Grosvenor & Renz* 1292 (K; SRGH). S: Lake Malawi, Boadzulu Island, 14.iii.1955, *Exell, Mendonça & Wild* 883 (BM; LISC). **Mozambique**. N: Erati, entre Nampula e Nacaroa, 30.x.1942, *Mendonça* 1141 (EA; K; LISC; PRE). Z: Alto Molocue, Gilé, ao 10 km. Monte Gilé, 21.xii.1967, *Torre & Correia* 16679 (LISC). T: Rio Mucangadzi, do 5 km. do barragem, proximo do posto Policial No. 3, Estrada nova para Meroeira Cahora Bassa, 30.i.1973, *Torre, Carvalho & Ladeira* 18935 (LISC). MS: Manica, Chimoio a 19 km. de Manica, 24.xi.1965, *Torre & Correia* 13226 (LISC; LMA). M: Maputo near Swaziland border, between Goba and Stegi, 2.vii.1948, *Dyer* 5054 (K; PRE).

Also known from Guinea to Ethiopia and the Somali Rep., and disjunctly through Tanzania to S. Africa (Natal). Rocky hillsides, rocky outcrops, banks of seasonally dry rivers, on granite, banded ironstone or sandstone; 0–1100 m.

The leaves of *F. abutilifolia* resemble those of *F. tettensis* but are larger (up to 17 × 18 cm. as opposed to 12 × 12 cm.) and are nearly glabrous.

20. **Ficus trichopoda** Bak. in Journ. Linn. Soc., Bot. **20**: 261 (1883). —K. Coates Palgrave, Trees Southern Africa: 117 (1977). —C.C. Berg et al. in Fl. Cameroun **28**: 158, t. 52 (1985). —C.C. Berg in F.T.E.A., Moraceae: 68 (1989); in Kirkia **13**: 265 (1990). —van Greuning in S. Afr. J. Bot. **56**, 6: 614, fig. 14 (1990). Type from Madagascar.

 Ficus congensis Engl., Bot. Jahrb. **8**: 59 (1886). —Hutch. in F.T.A. **6**, 2: 195 (1917). —Keay in F.W.T.A. ed. 2, **1**: 609 (1958). —F. White, F.F.N.R.: 33, t. 6L (1962). Type from Zaire.

 Ficus zuvalensis Sim, For. Fl. Port. E. Afr.: 100 t. 93A (1909). Type: Mozambique, Quissico (Cusico) *Sim* 5515, (K, holotype).

 Ficus hippopotami Gerstner in Journ. S. Afr. Bot. **9**: 151 (1943). Type from S. Africa.

Tree up to 10(20) m. tall, or a shrub, terrestrial, often with stout stilt- or pillar-roots. Leafy twigs 3–7 mm. in diam., glabrous or white puberulous to hirtellous, periderm not flaking off. Leaves spirally arranged; lamina ± broadly ovate to elliptic, 6–20(28) × 4–12(32) cm., coriaceous; apex shortly acuminate to obtuse; base obtuse to cordate; margin entire; superior surface glabrous, puberulous to hirtellous on the main veins, inferior surface white hirtellous to tomentellous at least on the midrib, sometimes glabrous; lateral veins 7–11 pairs, the basal pair branched, ending at the margin below or sometimes at the middle of the lamina, tertiary venation partly scalariform; petiole 2–4(7) cm. long, (1)2–3 mm. thick, epidermis not flaking off; stipules 1.5–4.5(8) cm. long, white puberulous to hirtellous, caducous. Figs up to 4 together in the leaf axils; peduncle 5–10 mm. long; basal bracts c. 2 mm. long, persistent. Receptacle subglobose, 10–20 mm. in diam. when fresh, 5–15 mm. in diam. when dry, glabrous or ± densely puberulous, smooth or verruculate, red to yellow at maturity.

Zambia. N: Mbala Distr., Isoko Valley, Mwambeshi R., 5.ix.1960, *Richards* 13192 (K; LISC). W: Solwezi Distr., Mutanda Bridge, 2.vii.1930, *Milne-Redhead* 647 (K). **Malawi**. C: Dedza Saw Mill, *Salubeni* 848 (SRGH). S: Zomba Distr., Malosa, Chilema Lay Training Centre, 31.x.1982, *Chapman* 6452 (K). **Mozambique**. N: andados 42 km. de Imala para Mocuburi, 16.i.1964, *Torre & Paiva* 10027 (COI; LISC; LMA: LMU; UC). Z: Chapala, S. of Alto Molócuè, ix.1971, *Bowbrick* J9/71 (LISC; SRGH). MS: Beira Distr., Cheringoma coast, Nyemesembe Fishing Camp, on Zuni Estuary, vii.1972, *Tinley* 2674 (K; PRE; SRGH). GI: Bazaruto Island, Jeckere Swamp, 7.xi.1958, *Mogg* 28934 (K; LISC; SRGH). M: Between Zitundo and Ponta do Ouro, 7.iv.1948, *Gomes e Sousa* 3723 (K).

Also known from Senegal to the Sudan, through Uganda and Tanzania to S. Africa (Natal), also in Madagascar. Mushitu (swamp forest), river banks, dambos and swamp grassland; 0–1200 m.

This species can be recognised by the relatively long stipules (1.5–4.5 cm. long, up to 8 cm. long on new flush).

21. **Ficus lutea** Vahl, Enum. Pl. **2**: 185 (1805). —Hutch. in F.T.A. **6**, 2: 215 (1917). —C.C. Berg in Kew Bull. **36**: 597 (1981); in Fl. Cameroun **28**: 206, t. 73 (1985); in F.T.E.A., Moraceae: 69 (1989); in Kirkia **13**: 261 (1990). —van Greuning in S. Afr. J. Bot. **56**, 6: 609, fig. 9 (1990). Neotype from Ghana, chosen by Berg in Kew Bull. **36**: 597 (1981).

 Urostigma luteum (Vahl) Miq. in Hook., Lond. Journ. Bot. **6**: 554 (1847).

 Ficus vogelii (Miq.) Miq. in Ann. Mus. Bot. Lugd.-Bat. **3**: 288 (1867). —Keay in F.W.T.A., ed. 2, **1**: 609 (1958). —K. Coates Palgrave, Trees Southern Africa: 118 (1977). Lectotype from Liberia, chosen by Berg in Fl. Gabon **26**: 193 (1984).

 Ficus quibeba Ficalho, Pl. Ut. Afr. Port.: 270 (1884). Type from Angola.

 Ficus subcalcarata Warb. & Schweinf. in Engl., Bot. Jahrb. **20**: 155 (1894). Type from Zaire.

 Ficus holstii Warb. in Engl., Bot. Jahrb. **20**: 160 (1894). Type from Tanzania.

 Ficus lanigera Warb. in Engl., Bot. Jahrb. **20**: 162 (1894). Type from Tanzania.

 Ficus verrucocarpa Warb. in Engl., Bot. Jahrb. **30**: 294 (1901). Type from Tanzania.

 Ficus cabrae Warb. in Ann. Mus. Congo, Bot. sér. 6, **1**: 9 (1904). Type from Zaire.

 Ficus nekbudu Warb. in Ann. Mus. Congo, Bot. sér. 6, **1**: 6 (1904). —Hutch. in F.C. **5**, 2: 535 (1925). —Lebrun & Boutique in F.C.B. **1**: 145 (1948). Type from Zaire.

 Ficus utilis Sim, For. Fl. Port. E. Afr.: 100, t. 91 (1909). Type: Mozambique, Maputo, *Sim* 6125 (K, holotype).

 Ficus akaie De Wild. in Bull. Soc. Roy. Bot. Belg. **52**: 198 (1913). Type from Zaire.

 Ficus incognita De Wild. in Bull. Soc. Roy. Belg. **52**: 213 (1913). Type from Zaire.

 Ficus kaba De Wild. in Bull. Soc. Roy. Belg. **52**: 213 (1913). Type from Zaire.

Tree up to 20(25) m. tall, hemi-epiphytic or secondarily terrestrial, with a spreading crown. Leafy twigs 5–12(20) mm. thick, puberulous, white to yellow tomentose, hirsute or subvillous, or glabrous, periderm flaking off when dry. Leaves spirally arranged; lamina elliptic to oblong or subobovate, 7–25(45) × 3–12(20) cm., coriaceous; apex acuminate; base obtuse, cuneate or subcordate; margin entire; superior surface glabrous, puberulous on the midrib, inferior surface puberulous to hirtellous or glabrous, the main veins ± tomentose-hirsute below; lateral veins (4)6–10(12) pairs, tertiary venation partly scalariform to reticulate; petiole 1.5–13(17) cm. long, 2–4(8) mm. thick, epidermis flaking off when dry; stipules 5–25 mm. long, up to 80 mm. long on new flush, puberulous or white- to yellow-subsericeous, caducous. Figs up to 4 together in the leaf axils or just below the leaves, sessile, initially enclosed in a white-pubescent to subhirsute calyptrate bud up to 0.5 cm. long; basal bracts 3–6(8) mm. long, persistent. Receptacle subglobose, c. 10–25(35) mm. in diam. when fresh, 8–15(25) mm. in diam. when dry, puberulous to white- or yellow-pubescent to subhirsute, smooth or verruculose, yellow to orange or brownish at maturity.

Zambia. N: Mbala Distr., Vomo Gap, Fwambo side, 27.ix.1960, *Richards* 13280 (K; LISC). W: Mushishima, Chingola, 30.v.1960, *Mutimushi* 2617 (K; NDO). **Zimbabwe.** N: Murewa (Mwera), 15.x.1926, *Eyles* 1199 (K). E: Chipinge Distr., Chirinda Forest Margin, vi.1964, *Goldsmith* 19/64 (K; LISC; PRE; SRGH). S: Bikita Distr., S. slope of Mt. Horzi on outer edge of ravine, 10.v.1969, *Pope* 147 (K; SRGH). **Malawi.** N: Nkhata Bay, 4.vi.1981, *White* (FHO). S: Mulanje Mt., Litchenya, 27.iii.1957, *Chapman* 349 (FHO; K). **Mozambique.** N: Mandimba Valley, 1.xi.1941, *Hornby* 2371 (PRE). Z: entre Muobede e Nhamarroi, 3.vi.1943, *Torre* 5422 (K). T: Zóbuè, no amo do monte Zóbuè, 3.x.1942, *Mendonça* 627 (LISC; PRE). MS: Inhambane Distr., Quissico, estrada para Chiducoane, 17.xii.1944, *Mendonça* 3301 (K). GI: Chidenguele margem da Lazoh, 18.viii.1947, *Pedro & Pedrógão* 1834 (PRE). M: Regulo de Banhanine, Manjacaze, estrada para Chidenguele, 19.iii.1948, *Torre* 7525 (K).

Widespread in subsaharan Africa, from the Cape Verde Islands to Ethiopia and southwards to Angola and S. Africa (Natal). Also in Madagascar, and the Seychelles, Aldabra and Comoro Islands. Evergreen and riverine forest; 0–1800 m.

22. **Ficus fischeri** Mildbr. & Burret in Engl., Bot. Jahrb. **46**: 227 (1911). —Hutch. in F.T.A. **6**, 2: 126 (1916). —F. White, F.F.N.R.: 34 (1962). —K. Coates Palgrave, Trees Southern Africa: 108 (1977). —C.C. Berg in F.T.E.A., Moraceae: 70 (1989); in Kirkia **13**: 265 (1990). —van Greuning in S. Afr. J. Bot. **56**, 6: 617, fig. 17 (1990). TAB. **20**. Lectotype from Tanzania, chosen by Berg in F.T.E.A., Moraceae: 70 (1989).

Ficus kiloneura Hornby in Bothalia **4**: 1007 (1948). Type: Mozambique, Niassa, near Chiponde Frontier Post, *Hornby* 2471 (PRE, holotype; K, isotype).

Tree up to 15 m. tall, hemi-epiphytic, soon terrestrial; crown flat-topped. Leafy twigs 4–10 mm. thick, glabrous or puberulous, periderm not flaking off. Leaves spirally arranged; lamina ovate to elliptic, (4)6.5–17 × (3)5–11 cm., coriaceous; apex acuminate, sometimes subacute or rounded, base cordate to truncate, sometimes rounded; margin entire; both surfaces glabrous; lateral veins 9–15 pairs, tertiary venation parallel to the lateral veins or reticulate; petiole 2.5–10 cm. long, (1)1.5–3 mm. thick; stipules 3–8 mm. long, glabrous or puberulous, caducous. Figs solitary or in pairs in the leaf axils; peduncle 8–18 mm. long; basal bracts 2–2.5 mm. long, caducous. Receptacle globose, 1.5–2 cm. in diam. when fresh, 1.5–2 cm. in diam. when dry, glabrous or minutely brownish puberulous; yellowish green at maturity; wall of fruiting fig c. 2 mm. thick when dry, ± wrinkled.

Botswana. N: Botswana side of Kwando mainstream, 18°06'S, 23°21'E, 27.vii.1973, *P.A. Smith* 686 (K; PRE; SRGH). **Zambia.** B: Nangweshi, near Zambezi R., 21.vii.1952, *Codd* 7134 (BM; K; PRE; SRGH). C: c. 40 km. from Lusaka on Kabwe road, 19.xi.1982, *Berg & Bingham* 1381 (K; U). S: Tungzi Area, Mulobezi, 14.x.1947, *Greenway & Brenan* 7948 (K; PRE). **Zimbabwe.** E: Chipinge, 24–32 km. S. of Mt. Selinda, xii.1972, *Burrows* 551 (SRGH). **Malawi.** S: Mulanje Distr., Phalombe–Jali road, Masamba Village, 19.viii.1983, *Seyani* 1220 (MAL). **Mozambique.** N: Maniamba, 100 km. a Boronango-Mecaloja, 13.ix.1934, *Torre* 576 (COI; LISC; PRE). Z: c. 3 km. E. of Alto Molócuè, 29.vi.1971, *van Niekerk* S2B (K; SRGH). MS: Buzi, Régulo Mucheve, 20.xi.1964, *Carvalho* 760 (K; LMA).

Also in Zaire (Shaba), Tanzania and Angola. Riverine vegetation, miombo woodland, *Baikiaea* woodland and wooded grassland, often on termitaria; 900–1500 m.

23. **Ficus craterostoma** Mildbr. & Burret in Engl., Bot. Jahrb. **46**: 247 (1911). —Hutch. in F.T.A. **6**, 2: 160 (1916); in F.C. **5**, 2: 536 (1925). —Lebrun & Boutique in F.C.B. **1**: 148 (1948). —F. White, F.F.N.R.: 32, t. 6G (1962). —K. Coates Palgrave, Trees Southern Africa: 107 (1977). —C.C. Berg et al. in Fl. Cameroun **28**: 182, t. 62 (1985). —C.C. Berg in F.T.E.A., Moraceae: 71 (1989); in Kirkia **13**: 266 (1990). —van Greuning in S. Afr. J. Bot. **56**, 6: 616, fig. 16 (1990). Lectotype from Tanzania, chosen by Berg in F.T.E.A., Moraceae: 71 (1989).

Ficus luteola De Wild. in Fedde, Repert. **12**: 199 (1913). —Hutch. in F.T.A. **6**, 2: 159 (1916). —Lebrun & Boutique in F.C.B. **1**: 142 (1948). Type from Zaire.

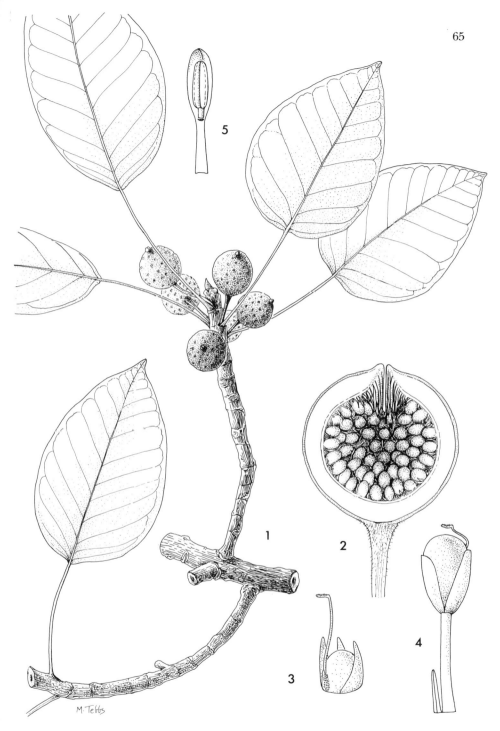

Tab. 20. FICUS FISCHERI. 1, fertile twig (×⅔), *Berg & Bingham* 1381; 2, l/s through fig (× 3), *Menezes* 3555; 3, diagram of long-styled pistillate flower, with 3 free or basally connate tepals and long style (× 12); 4, diagram of short-styled pistillate flower (× 14); 5, diagram of staminate flower, showing stamen enclosed within 3 tepals, 1 tepal removed (× 14), 3–5 *Berg* 1381. Drawn by M. Tebbs & S. Dawson.

Ficus pilosula De Wild. in loc. cit. Type from Zaire.
Ficus rubropunctata De Wild. in loc. cit. Type from Zaire.
Ficus furcata var. *angustifolia* De Wild. in tom. cit.: 303 (1913). Type from Zaire.
Ficus anomani Hutch. in Bull. Misc. Inf., Kew **1915**: 331 (1915). —Keay in F.W.T.A., ed. 2, **1**:
607 (1958). Lectotype from Ghana, chosen by Berg in Fl. Gabon **26**: 170 (1984).
Ficus mutantifolia Hutch. in Bull. Misc. Inf., Kew **1915**: 330 (1915). Type from Angola.
Ficus ruwenzoriensis De Wild. in Ann. Soc. Sci. Brux. **40**: 282 (1921). Type from Zaire.

Tree up to 10 m. tall, or a shrub, hemi-epiphytic. Leafy twigs 2–5 mm. thick, glabrous or white-puberulous to hirtellous, periderm sometimes flaking off when dry. Leaves spirally arranged, tending to be distichous, often subopposite; lamina narrowly obtriangular to subobovate, or oblong to elliptic, 3–8 × 2–4.5 cm., subcoriaceous; apex truncate to emarginate (bilobed) or obtuse; base acute to obtuse; margin entire; both surfaces glabrous; lateral veins 5–10 pairs, midrib not reaching the apex of the lamina, tertiary venation reticulate or tending to run parallel to the lateral veins; petiole 5–20 mm. long, 1–2 mm. thick; stipules c. 5 mm. long, glabrous or yellowish-whitish puberulous, subpersistent or caducous. Figs in pairs in the leaf axils, sessile, initially enclosed in a calyptrate bud up to 1 cm. long which later splits into two subpersistent or caducous parts, pubescent inside; basal bracts 1–1.5 mm. long, persistent. Receptacle globose to ellipsoid, 8–12 mm. in diam. when fresh, c. 5 mm. in diam. when dry, glabrous or puberulous, reddish (or yellowish) at maturity; wall slightly wrinkled when dry.

Zambia. N: Mansa Distr., edge of Lake Bangweulu, near Samfya Mission, 20.viii.1952, *White* 3091 (BM; COI; FHO; K; PRE). W: Solwezi Distr., Kifubwa (Chifubwa) R. Gorge, 3 km. S. of Solwezi, 20.iii.1961, *Drummond & Rutherford-Smith* 7125 (K; NDO; PRE; SRGH). **Zimbabwe**. E: Mutare Distr., Vumba Mt., near Cloudlands, 25.v.1961, *Plowes* 2184 (K; SRGH). **Mozambique**. MS: Manica Distr., Bárùe, Serra de Choa, andado 9 km. de Catandica (Vila Gouveia), 24.v.1971, *Torre & Correia* 18613 (LISC; LMA; LMU).
Also in Sierra Leone eastwards to Uganda, and southwards to Angola and S. Africa (NE. Cape Province). Swamp forest (mushitu) and evergreen forest margins, riverine and lake shore vegetation, often on deep sand; 0–2100 m.
This species can be confused with *F. natalensis* and *F. thonningii*. It can be distinguished from the former by its sessile figs and from the latter by its subopposite leaves and the calyptrate buds up to c. 10 mm. long with subpersistent scales.

24. **Ficus lingua** De Wild. & T. Durand in Ann. Mus. Congo, Bot., sér. 3, **2**: 216 (1901). —Warb. in Ann. Mus. Congo, Bot., sér. 6, **1**: 24 (1904). —Hutch. in F.T.A. **6**, 2: 156 (1916). —Lebrun & Boutique in F.C.B. **1**: 138 (1948). —Keay in F.W.T.A., ed. 2, **1**: 608 (1958). —C.C. Berg et al. in Fl. Cameroun **28**: 180, t. 61 (1985). —C.C. Berg in Kew Bull. **43**: 85 (1988); in F.T.E.A., Moraceae: 72 (1989); in Kirkia **13**: 266 (1990). Type from Zaire.
Ficus depauperata Sim, For. Fl. Port. E. Afr.: 98, t. 90B (1909). —Hutch. in F.T.A. **6**, 2: 204 (1917); in F.C. **5**, 2: 538 (1925). —K. Coates Palgrave, Trees Southern Africa: 107 (1977). Type: Mozambique, Maputo, Katembe, *Sim* 5031 (K, holotype).

Tree up to 30 m. tall, hemi-epiphytic, secondarily terrestrial. Leafy twigs 1–4 mm. thick, whitish puberulous, periderm of older plants flaking off when dry. Leaves spirally arranged, tending to be distichous; lamina oblanceolate to subobovate or narrowly obtriangular, 0.5–5 × 0.3–2(3) cm., subcoriaceous; apex acute to truncate or emarginate; base cuneate to obtuse; margin entire; both surfaces glabrous; lateral veins 5–8 pairs, tertiary venation reticulate or tending to run parallel to the lateral veins; petiole 2–8 mm. long, 0.5–1 mm. thick; stipules 2–5 mm. long, puberulous or ciliolate only, caducous, sometimes subpersistent. Figs in pairs in the leaf axils or just below the leaves, on peduncles 2–5 mm. long, sometimes subsessile; basal bracts 1–2 mm. long, persistent. Receptacle globose, sometimes ellipsoid, c. 5 mm. in diam. when fresh, c. 3–4 mm. in diam. when dry, minutely puberulous, reddish or yellowish at maturity; wall of fruiting fig thin, smooth or slightly wrinkled when dry.

Malawi. S: Nsanje, Boma near Police Station, 12.vii.1960, *Willan* 48 (FHO; LISC; MAL; PRE; SRGH). **Mozambique**. N: Cabo Delgado, entre Macomia e Mipande, 30.ix.1948, *Barbosa* 2301 (COI; EA; LISC; PRE; WAG). MS: Maringue, v.1973, *Bond* 9B 66 (LISC; SRGH). GI: Vilanculos, Mapinhane, 31.viii.1942, *Mendonça* 50 (COI; EA; FHO; LISC; LMU; PRE; SRGH).
Also in Kenya and Tanzania. Coastal bushland and forest; 0–1200 m.

25. **Ficus natalensis** Hochst. in Flora **28**: 88 (1845). —Hutch. in F.T.A. **6**, 2: 208 (1917); in F.C. **5**, 2: 538 (1925). —Keay in F.W.T.A. ed. 2, **1**: 610 (1958). —F. White, F.F.N.R.: 34 (1962). —K. Coates Palgrave, Trees Southern Africa: 110 (1977). —C.C. Berg et al. in Fl. Cameroun **28**: 184, t. 63 (1985). —C.C. Berg in Kew Bull. **43**: 87 (1988); in F.T.E.A., Moraceae: 71 (1989); in Kirkia **13**: 266

(1990). Lectotype from S. Africa, chosen by Berg in F.T.E.A., Moraceae: 71 (1989).
    Urostigma natalense (Hochst.) Miq. in Hook., Lond. Journ. Bot. **6**: 556 (1847).

Tree up to 30 m. tall, or a shrub, hemi-epiphytic or secondarily terrestrial, sometimes semi-scandent. Leafy twigs 2–5 mm. thick, glabrous or sparsely minutely puberulous, periderm not flaking off. Leaves spirally arranged, or ± distichous, often subopposite; lamina oblong to elliptic or obovate to broadly obtriangular, occasionally lanceolate, 2.5–10 × 1–4.5 cm., subcoriaceous; apex shortly acuminate to obtuse or subacute to rounded or emarginate; base acute to obtuse; margin entire; both surfaces glabrous; lateral veins 6–13 pairs, midrib usually not reaching the apex of the lamina, tertiary venation reticulate or ± parallel to the lateral veins; petiole 5–20(30) mm. long, 1–2(2.5) mm. thick, glabrous; stipules 2–10 mm. long, glabrous or puberulous, caducous. Figs in pairs in the leaf axils or sometimes also just below the leaves, initially enclosed by a calyptrate bud cover, up to 1.5 cm. long, subovoid and ± glabrous; peduncle 2–10 mm. long; basal bracts 1.5–2.5 mm. long, caducous sometimes subpersistent. Receptacle often shortly stipitate, at least when dry, globose to ellipsoid or obovoid, c. 1.5–2 cm. in diam. when fresh, 0.8–1.5 cm. in diam. when dry, glabrous, reddish-orange or yellowish (to brown) at maturity; wall (rather thin) usually wrinkled when dry, apex plane or slightly protruding.

Subsp. **natalensis** —C.C. Berg in Kew Bull. **43**: 87 (1988); in Kirkia **13**: 266 (1990). —van Greuning in
    S. Afr. J. Bot. **56**, 6: 620, fig. 19 (1990).
    Ficus volkensii Warb. in Engl., Bot. Jahrb. **20**: 167 (1894). —Hutch. in F.T.A. .**6**, 2: 208 (1917).
    Type from Tanzania.
    Ficus natalensis var. *latifolia* Warb. in Viert. Nat. Ges. Zürich **51**: 142 (1906). Type from S.
    Africa.
    Ficus durbanii Warb. in Viert. Nat. Ges. Zürich **51**: 142 (1906). Type from S. Africa.
    Ficus natalensis var. *pedunculata* Sim, For. Fl. Port. E. Afr.: 98, t. 90A (1909). Type:
    Mozambique, Maputo, (Lourenço Marques) *Sim* 5729 (K, holotype).
    Ficus variabilis forma *obtusifolia* De Wild. in Ann. Soc. Sci. Brux. **40**: 282 (1921). Type from
    Zaire.
    Ficus variabilis forma *subacuminata* De Wild. in Ann. Soc. Sci. Brux. **40**: 283 (1921). Type from
    Zaire.
    Ficus scutata Lebrun in Mém. Inst. Roy. Colon. Belge, sect. Sci. Nat. 8°, 3, 1: 68, t. 4B (1934).
    Type as for *F. variabilis* forma *obtusifolia* De Wild.
    Ficus subacuminata (De Wild.) Lebrun in Mém. Inst. Roy. Colon. Belge, sect. Sci. Nat. 8°, 3, 1:
    70, t. 4C (1934).

Tree or shrub. Leafy twigs glabrous or sparsely puberulous. Leaf lamina oblong to elliptic or obovate, sometimes broadly so or obtriangular; apex acuminate to rounded or emarginate; margin flat, entire, midrib usually terminating near the leaf apex, lateral veins 6–13 pairs; stipules glabrous or sparsely minutely puberulous. Calyptrate bud cover enclosing the young figs, often conspicuous, up to 1 cm. long. Figs 0.8–1.5 cm. in diam. when dry; basal bracts caducous.

Zambia. N: Mansa, in grounds of Samfya Mission School, Lake Bangweulu, 23.viii.1952, *Angus* 326 (BM; FHO). W: Matonchi Farm, Matonchi R., 29.i.1938, *Milne-Redhead* 4396 (K; PRE). **Zimbabwe.** N: Bindura Distr., Chindamora Reserve, 56 km. N. of Harare, 21.ix.1959, *Goodier* 612 (K; PRE; SRGH). W: Matopos Distr., Moth Shrine Koppie, 23.xi.1951, *Plowes* 1332 (K; SRGH). C: Makoni Distr., Rusape, 25.v.1952, *Dehn* in GHS 96170 (K; PRE; SRGH). E: Mutare Distr., NW. slope of Murahwa's Hill, 10.vii.1963, *Chase* 8042 (K; PRE; SRGH). S: Mberengwa Distr., Buhwa Mt., main ridge, 30.x.1973, *Pope, Biegel & Gosden* 1124 (PRE; SRGH). **Malawi.** N: Nkhata Bay Distr., c. 18km. S. of Nkhata Bay Junction at Sanga, 23.v.1976, *Pawek* 11315 (K; MO; PRE; UC). S: Mulanje Mt., 11.vi.1957, *Chapman* 381 (BM; FHO; K; PRE). **Mozambique.** Z: Maganja da Costa, ao km. 26 de Vila da Maganja estrada para Bajone, 23.xi.1967, *Torre & Correia* 16224 (K; LISC). MS: Beira, Chiniziua, near the village Zuni, 10.iv.1957, *Gomes e Sousa* 4378 (COI; K; PRE). GI: Macia, Incaia, 17.vii.1947, *Pedro & Pedrogão* 1489 (COI; K; LMA). M: Jardim Tunduru (Vasco da Gama), 1.iv.1971, *Balsinhas* 1821 (K; LMA; LISC).

    Also known from Kenya, Uganda, Zaire (Oriental and Shaba) and S. Africa (Natal). Evergreen, riverine and coastal forest, *Brachystegia/Uapaca* and miombo woodland and submontane grassland, often in rocky places; 0–1700 m.

Subsp. **leprieurii** (Miq.) C.C. Berg in Kew Bull. **43**, 1: 88 (1988); in Kirkia **13**: 267 (1990). Type from
    "Senegambia".
    Ficus leprieurii Miq. in Ann. Mus. Bot. Lugd.-Bat. **3**: 219 (1867). —Hutch. in F.T.A. **6**, 2: 158
    (1916). —Lebrun & Boutique in F.C.B. **1**: 139, t. 15 (1948). —Keay in F.W.T.A. ed. 2, **1**: 608 (1958).
    Type as above.

Ficus chrysocerasus Warb. in Engl., Bot. Jahrb. **20**: 167 (1894). Type from Angola.
Ficus furcata Warb. in Engl., Bot. Jahrb. **20**: 173 (1894). Type from Zaire.
Ficus leprieurii var. *sessilis* Hutch. in F.T.A. **6**, 2: 159 (1916). Type as for *F. furcata* Warb.
Ficus brevipedicellata De Wild. in Ann. Soc. Sci. Brux. **40**: 279 (1921). Type from Zaire.

Tree or shrub, often semi-scandent. Leafy twigs glabrous, minutely puberulous or white to yellowish hirtellous. Leaf lamina elliptic to oblong, or broadly obovate to obtriangular; apex acuminate to rounded, emarginate or truncate; margin often revolute; midrib terminating at the leaf apex or well before it. Lateral veins (4)6–10 pairs; stipules usually subsericeous to hirtellous with white to yellowish hairs, at least at the base, occasionally glabrous or sparsely puberulous, caducous leaving a fimbriate scar. Calyptrate bud cover enclosing minute young figs. Figs 0.5–1 cm. in diam. when dry; basal bracts caducous sometimes subpersistent.

Zambia. W: Mwinilunga Distr., c. 12 km. W. of Kakoma, 19.ix.1952, *White* 3413 (FHO; K).
Also known from Senegal to S. Sudan and southwards to E. Zaire and to N. Angola. Forest, often along rivers and streams; 0–1200 m.

26. **Ficus burtt-davyi** Hutch. in Bull. Misc. Inf., Kew **1916**: 232 (1916); in F.C. **5**, 2: 540 (1925). —K. Coates Palgrave, Trees Southern Africa: 104 (1977). —C.C. Berg in Kirkia **13**: 267 (1990). —van Greuning in S. Afr. J. Bot. **56**, 6: 614, fig. 15 (1990). Lectotype from S. Africa, chosen by C.C. Berg.
Urostigma natalense var. *minor* Sond. in Linnaea **23**: 137 (1850). Lectotype from S. Africa, chosen by Berg in Fl. Gabon **26**: 173 (1984).
Ficus natalensis var. *minor* (Sond.) Warb. in Viert. Nat. Ges. Zürich **51**: 142 (1906).
Ficus natalensis var. *puberula* Warb. in Viert. Nat. Ges. Zürich **51**: 142 (1906). Type from S. Africa.

Tree up to 8 m. tall, a scrambling shrub or ± lianescent, terrestrial or epilithic. Leafy twigs 1.5–3 mm. thick, puberulous, periderm not flaking off. Leaves spirally arranged, or ± distichous; lamina elliptic to obovate, sometimes oblong, ovate or subcircular, (0.5)1.5–10 × (0.3)0.7–4(5.5) cm., coriaceous; apex shortly acuminate to subacute or obtuse; base obtuse to rounded or emarginate; margin entire, revolute towards the base; both surfaces glabrous, or inferior surface sparsely minutely puberulous on the midrib; lateral veins (3)5–6(7) pairs, the basal pair unbranched, midrib reaching the apex of the lamina, tertiary venation reticulate; petiole 4–15(27) mm. long, c. 1 mm. thick; stipules 2–8 mm. long, to 5 cm. long on new flush, ciliolate, caducous. Figs in pairs or solitary in the leaf axils; peduncle 1–4(7) mm. long; basal bracts 1–1.5 mm. long, persistent. Receptacle globose, 5–12 mm. in diam. when dry; sparsely minutely puberulous, wall thin, dark brown to black at maturity.

Mozambique. M: Maputo, Ponta do Ouro, 18.xi.1944, *Mendonça* 2909 (LISC).
Also in S. Africa (Natal, Cape Province). Coastal and dune forest, or scrub, sometimes in coastal swamp forest.

27. **Ficus thonningii** Blume in Rumphia **2**: 17 (1836). —Hutch. in F.T.A. **6**, 2: 187 (1916). —Lebrun & Boutique in F.C.B. **1**: 148. t. 16 (1948). —Keay in F.W.T.A. ed. 2, **1**: 610 (1958). —C.C. Berg et al. in Fl. Cameroun **28**: 175, t. 59 (1985). —C.C. Berg in F.T.E.A., Moraceae: 73, t. 21 (1989); in Kirkia **13**: 268 (1990). —van Greuning in S. Afr. J. Bot. **56**, 6: 622, fig. 20 (1990). Type from Ghana.
Ficus microcarpa Vahl, Enum. Pl. **2**: 188 (1805) non L.f. (1781) nom. illegit. Type as above.
Urostigma burkei Miq. in Hook., Lond. Journ. Bot. **6**: 555 (1847). Type from S. Africa.
Urostigma thonningii (Blume) Miq. in Hook., Lond. Journ. Bot. **6**: 558 (1847).
Ficus burkei (Miq.) Miq. in Ann. Mus. Bot. Lugd.-Bat. **3**: 289 (1867). —Hutch. in F.T.A. **6**, 2: 202 (1917). —F. White, F.F.N.R.: 34 (1962).
Ficus psilopoga Ficalho, Pl. Ut. Afr. Port.: 270 (1884). Type from Angola.
Ficus persicifolia Warb. in Engl., Bot. Jahrb. **20**: 162 (1894). —F. White, F.F.N.R.: 33 (1962). Type from Angola.
Ficus chlamydodora Warb. in Engl., Bot. Jahrb. **20**: 163 (1894). Type from Tanzania.
Ficus petersii Warb. in Engl., Bot. Jahrb. **20**: 164 (1894). Type: Mozambique, Moravi, *Peters* s.n. (B, holotype).
Ficus mabifolia Warb. in Engl., Bot. Jahrb. **20**: 165 (1894). Type from Tanzania.
Ficus medullaris Warb. in Engl., Bot. Jahrb. **20**: 169 (1894). Type from Zaire.
Ficus goetzei Warb. in Engl., Bot. Jahrb. **28**: 378 (1900). Type from Tanzania.
Ficus ruficeps Warb. in Engl., Bot. Jahrb. **30**: 294 (1901). Type from Tanzania.
Ficus eriocarpa Warb. in Engl., Bot. Jahrb. **30**: 294 (1901). Type from Tanzania.
Ficus persicifolia var. *angustifolia* Warb. in Ann. Mus. Congo, Bot. sér. 6, **1**: 15 (1904). Type from Zaire.
Ficus persicifolia var. *glabripes* Warb. in Ann. Mus. Congo, Bot. sér. 6, **1**: 15 (1904). Syntypes from Zaire.

Ficus pubicosta Warb. in Ann. Mus. Congo, Bot. sér. 6, **1**: 16 (1904). Type from Zaire.
Ficus galpinii Warb. in Viert. Nat. Ges. Zürich **51**: 140 (1906). Lectotype from S. Africa, chosen by Berg in Fl. Gabon **26**: 162 (1984).
Ficus dinteri Warb. in Viert. Nat. Ges. Zürich **51**: 141 (1906). Type from Namibia.
Ficus schinziana Warb. in Viert. Nat. Ges. Zürich **51**: 143 (1906). Type from S. Africa.
Ficus rhodesiaca Mildbr. & Burret in Engl., Bot. Jahrb. **46**: 254 (1911). Type: Zimbabwe, Harare, Engler 3060 (B, holotype; K, isotype).
Ficus cyphocarpa Mildbr. in Engl., Bot. Jahrb. **46**: 261 (1911). Type from Zaire.
Ficus bequaertii de Wild. in Bull. Soc. Roy. Bot. Belge **52**: 200 (1914). Type from Zaire.
Ficus erici-rosenii R.E. Fries, Wiss. Ergebn. Schwed. Rhod.-Kongo-Exped. 1911–1912 **1**: 15, t. 1 (1914). Type: Zimbabwe, Victoria Falls, Fries 30 (K, isotype).
Ficus butaguensis De Wild. in Ann. Soc. Sci. Brux. **40**: 279 (1921). Type from Zaire.
Ficus cognata N.E. Brown in Bull. Misc. Inf., Kew **1921**: 297 (1921). Type from Zaire.
Ficus crassipedicellata De Wild., Pl. Bequaert. **1**: 335 (1922). Type not designated.
Ficus crassipedicellata forma angustifolia De Wild., Pl. Bequaert. **1**: 336 (1922). Type from Zaire.
Ficus crassipedicellata forma boonei De Wild., Pl. Bequaert. **1**: 337 (1922). Type from Zaire.
Ficus crassipedicellata var. cuneata De Wild., Pl. Bequaert. **1**: 338 (1922). Type from Zaire.
Ficus mammigera R.E. Fries in Notizbl. Bot. Gart. Berl. **8**: 669 (1924). Type from Kenya.
Ficus phillipsii Burtt Davy & Hutch. in Burtt Davy, Fl. Pl. Ferns Transv., part II: 442 (1926). Type from S. Africa.
Ficus dekdekena var. angustifolia Peter in Fedde, Repert., Beih. **40**, 2, 2: 106 (1932). Type from Tanzania.
Ficus thonningii var. heterophylla Peter in Fedde, Repert., Beih. **40**, 2, 2: 109. Descr.: 9, t. II. 2, 2 (1932). Types from Tanzania.
Ficus kagerensis Lebrun & Toussaint in Expl. Parc Nat. Kagera **1**: 41 (1948). Type from Rwanda.
Ficus neurocarpa Lebrun & Toussaint in Expl. Parc Nat. Kagera **1**: 42 (1948). Type from Rwanda.
Ficus rupicola Lebrun & Toussaint in Expl. Parc Nat. Kagera **1**: 43 (1948). Type from Rwanda.
Ficus trophyton Lebrun & Toussaint in Expl. Parc Nat. Kagera **1**: 43 (1948).

Tree up to 15(30) m. tall, or a shrub, terrestrial or hemi-epiphytic. Leafy twigs 1.5–8 mm. thick, minutely puberulous to hirtellous or white- to brown-pubescent, or glabrous on the stipule scars, sometimes entirely glabrous, periderm usually not flaking off. Leaves spirally arranged, occasionally subopposite; lamina elliptic to oblanceolate or subobovate to subovate, (1.5)3–12(18) × (1)1.5–6(9) cm., subcoriaceous; apex acuminate to obtuse or rounded; base cuneate to rounded or subcordate, often slightly inaequilateral; margin entire; superior surface glabrous or sparsely puberulous to pubescent, the midrib more densely so, the inferior surface glabrous or sparsely to densely white- to brownish-puberulous or pubescent on the whole surface, the main veins or only the midrib; lateral veins (5)7–12(16) pairs, midrib often reaching the apex of the lamina (even in leaves with a rounded apex), tertiary venation reticulate or parallel to the lateral veins; petiole (0.5)1–4(6) cm. long, 1–2 mm. thick, often (not depending on the size of the lamina or the position of the leaf on the twig) variable in length on the same twig, glabrous or puberulous, hirtellous or pubescent; stipules 3–10(20) mm. long, white to brown pubescent, puberulous or only ciliolate, caducous or subpersistent. Figs in pairs in the leaf axils, sometimes also below the leaves, sessile or on peduncles up to 1 cm. long; basal bracts 2–4 mm. long, persistent. Receptacle globose to ellipsoid, c. 5–10(12) mm. in diam. when fresh, 4–12(17) mm. in diam. when dry, glabrous or sparsely to densely white to brown puberulous or pubescent, reddish, yellowish or brownish at maturity; wall thin, mostly smooth or slightly wrinkled when dry; apex plane to strongly protruding when dry.

Botswana. N: Gubatsa Hills, at base of NE. facing cliff, 24.x.1972, Biegel, Pope & Russell 4050 (PRE). SE: Gaborone, Aedume Park, 22.x.1977, O.J. Hansen 3244 (K; PRE; SRGH). **Zambia**. B: Sesheke Distr., Bombwe Forest, 3.i.1953, Angus 1099 (FHO; K). N: Mbala Distr., Lumi R. Marsh, near Kawimbe, 12.x.1956, Richards 6430 (K). W: Solwezi, 22.vii.1964, Fanshawe 8826 (K; LISC; NDO). C: Lusaka, Chelston School, 4.xii.1976, Bingham 2079 (K). E: Lundazi, 17.x.1967, Mutimushi 2240 (NDO). S: Mazabuka Distr., Siamambo For. Res., near Choma, 27.vii.1952, Angus 87 (FHO; K). **Zimbabwe**. N: Hillymead Farm, between Glendale & Bindura, 14.xii.1978, Arkell 2 (PRE; SRGH). W: Matopos Distr., Ntunja Cave, near Boomerang Farm, 7.ix.1952, Plowes 1467 (K; SRGH). C: Gweru Distr., Lower Gweru T.T.L., Chief Sogwala's house, 21.vii.1966, Biegel 1281 (K; SRGH). E: Chipinge, main road between Chipinge & Mt. Selinda, ii.1962, Goldsmith 53/62 (LISC; PRE; SRGH). S: Chibi Distr., near Madzivire Dip, c. 6.4 km. W. of Runde (Lundi) R. Bridge, 3.v.1962, Drummond 7907 (K; LISC; PRE; SRGH). **Malawi**. N: Mzimba Distr., Lundazi-Mzimba road, Mile 29, 28.iv.1952, White 2506 (FHO; K). C: Lilongwe Distr., c. 38 km. S. of Lilongwe, 6.iv.1978, Pawek 14326B (K; MAL; MO). S: Mt. Mulanje, path to Chambe Basin, 6.ix.1970, Müller 1556 (K; SRGH). **Mozambique**. N: Old road of Mandimba, Hornby 2344 (PRE). Z: Zambézia, Milange, 13.x.1942, Torre 4599 (LISC). T: Tete, Furancungo, 25.viii.1941, Torre 3328 (LISC). MS: Manica, Macequece

(Maciquece), 29.i.1948, *Mendonça* 3574 (LISC). M: Maputo Distr., Namaacha, Mt. Pondúini, 25.vii.1980, *Schäfer* 7199 (K; LMU).

Widespread in subsaharan Africa from Cape Verde Islands to Ethiopia, and southwards to Namibia and S. Africa (Cape Province). Also on the Islands of Pagolu, Príncipe, São Tomé and Bioko. Miombo woodland, swamp forest (mushitu), riverine vegetation, *Baikiaea* thickets (mukusi, mutemwa) on Kalahari Sand, wooded grassland and dambos, often on termitaria; 0–2300 m.

Ficus thonningii s.l. as here circumscribed is an extremely variable species or species complex. It cannot be satisfactorily subdivided on morphological criteria as one more or less distinct regional or ecological form gradually intergrades with other forms. However, two or more morphologically recognisable forms can occur side by side without the presence of intermediate specimens, which suggests the occurrence of reproductive isolation.

The recognition of infra-specific taxa might prove possible after further study of *F. thonningii* s.l. For the time being several forms representing the extremes of variation, and which are more or less clearly associated with geography and habitat, may be recognised.

The "**burkei**" form, is found in the savanna woodlands of the southern part of the species range. It usually has whitish (to pale brown) pubescent, pedunculate figs, and whitish pubescent twigs, petioles and/or leaf inferior surfaces. This form passes more or less gradually into forms in Mozambique, Angola, N. Zambia) with subsessile figs which are either glabrous or densely brown-pubescent. These latter forms have narrower (oblong to obovate or oblanceolate), often drooping leaves with relatively long petioles.

The "**mammigera**" form, found in the evergreen forests of E. Zimbabwe, Malawi and W. Zambia is almost glabrous and has relatively large, oblong to elliptic leaves and rather large, subsessile, often mammilate (when dry) figs.

28. **Ficus ottoniifolia** (Miq.) Miq. in Ann. Mus. Bot. Lugd.-Bat. **3**: 288 (1867). —Hutch. in F.T.A. **6**, 2: 134 (1916). —Lebrun & Boutique in F.C.B. **1**: 134 (1948). —Keay in F.W.T.A. ed. 2, **1**: 611 (1958). —C.C. Berg et al. in Fl. Cameroun **28**: 217, t. 77 (1985). —C.C. Berg in F.T.E.A., Moraceae: 76 (1989); in Kirkia **13**: 273 (1990). Type from Bioko.

Urostigma ottoniifolium Miq. in Hook., Lond. Journ. Bot. **6**: 557 (1847); **7**: 536, t. 13B (1848) nomen; in Hook., Niger Fl.: 521 (1849).

Tree up to 15 m. tall, or sometimes a shrub or liana, often terrestrial or sometimes hemi-epiphytic. Leafy twigs 2–5(10) mm. thick, minutely puberulous or glabrous, periderm not flaking off. Leaves arranged in spirals; lamina elliptic to oblong or subovate to subobovate, 6.5–15(22) × 3–7(9) cm., subcoriaceous; apex acuminate to subcaudate; base cuneate to rounded or subcordate; margin entire; both surfaces glabrous; lateral veins (4)6–12(14) pairs, tertiary venation reticulate; petiole (0.8)1.5–7 cm. long, 1.5–2 mm. thick; stipules 2–8 mm. long, up to c. 4 cm. long on new flush, sparsely puberulous or glabrous, caducous. Figs up to 4(10) together on spurs up to 1.5(4) cm. long on the older wood; bud scales of the spurs ± glabrous; peduncle 8–25 mm. long; basal bracts 2–3 mm. long, caducous or persistent. Receptacle ellipsoid to subglobose, 1.5–4 cm. in diam. when fresh, and 0.8–2.5 cm. in diam. when dry, puberulous or almost glabrous, greenish to pale orange or brownish at maturity, with pale green to whitish spots; wall c. 1 mm. thick when dry, not or hardly wrinkled.

Subsp. **macrosyce** C.C. Berg in Kew Bull. **43**: 90, fig. 4 (1988); in F.T.E.A., Moraceae: 77 (1989); in Kirkia **13**: 274 (1990). Type: Zambia, 5 km. S. of Solwezi, *Berg & Bingham* 1421 (BG; K; SRGH; WAG, isotypes; U, holotype).

Liana or small tree. Leaf lamina (4)8–16 × (1.2)3–7 cm., usually broadest in the middle, chartaceous; apex short- to long-acuminate, sometimes rounded; base emarginate to subcordate; lateral veins (4)6–8 pairs; petiole (0.8)1.5–7 cm. long, 0.5–1.5 mm. thick when dry. Peduncle 10–15(20) mm. long, c. 2 mm. thick when dry; basal bracts caducous. Receptacle ellipsoid to subglobose, 2.5–4 cm. in diam. when fresh, 2–2.5 cm. in diam. when dry; dark olive-green or red-brown, with pale green to white spots at maturity.

Zambia. B: Zambezi (Balovale), Chavuma, Zambezi R., 13.x.1952, *White* 3496 (BM; FHO; K). W: Kifubwa River Gorge, c. 5 km. S. of Solwezi, 26.xi.1982, *Berg & Bingham* 1421 (K; U).

Also in N. Namibia, Zaire (Shaba) and NE. Angola. Riverine forest, in rocky gorges and on rocks in rapids, also in swamp forest (mushitu).

This subspecies is usually a liana and differs from the other subspecies in having larger figs (when dry 2–2.5 cm. as compared with 0.7–1.8 cm. in diam.).

Subsp. **ulugurensis** (Mildbr. & Burret) C.C. Berg in Kew Bull. **43**: 92 (1988); in F.T.E.A., Moraceae: 77 (1989); in Kirkia **13**: 274 (1990). Lectotype from Tanzania, chosen by Berg in F.T.E.A.

Ficus ulugurensis Mildbr. & Burret in Engl., Bot. Jahrb. **46**: 226, t. 4 (1911). —Hutch. in F.T.A. **6**, 2: 139 (1916). Type as above.

Ficus scheffleri Mildbr. & Burret in Engl., Bot. Jahrb. **46**: 225 (1911). Type from Tanzania.
Ficus modesta F. White in Bull. Jard. Bot. Nat. Belg. **60**: 104 (1990). Type: Malawi, Chipalumbe gorge, (Mt. Mulanje, Phalombe R.), 13.v.1958, *Chapman* 567 (K, holotype; FHO, isotype).

Tree up to 10 m. tall, or a shrub. Leafy twigs dark-brown to blackish when dry. Leaf lamina 7.5–15 × 3–7 cm., subcoriaceous; apex shortly acuminate; base cuneate to rounded or subcordate; lateral veins 8–12(14) pairs; petiole 1.5–4 (5.5) cm. long, 1–1.5 mm. thick when dry. Peduncle basal bracts subpersistent. Receptacle ellipsoid to subglobose, 2–2.5 cm. in diam. when fresh, (1)1.2–1.8 cm. in diam. when dry, yellow with paler spots at maturity.

Malawi. C: Lilongwe, Nature Sanctuary, 5.ii.1985, *Patel & Banda* 2066 (MAL). S: Mt. Mulanje, Esperanza Tea Estate, 23.vii.1973, *Dowsett-Lemaire* 864 (FHO); Machemba Hill, 3.ix.1983, *Seyani & Balaka* 1312 (MAL).
Also recorded from Uganda, Kenya and Tanzania. In medium altitude rainforest, dry forest or rocky slopes and in transition woodland below evergreen forest; 850–1200 m.
This subspecies differs from the other subspecies in having ± persistant basal bracts on the peduncle, and leafy twigs drying dark brown instead of pale brown.
The other subspecies and their ranges of distribution are as follows; subsp. *ottoniifolia* ranges from Uganda to Sierra Leone (or Senegal?), subsp. *multinervia* C.C. Berg from Sierra Leone to Ivory Coast, and subsp. *lucanda* (Ficalho) C.C. Berg from NW. Tanzania through Zaire to Angola and Gabon.

29. **Ficus tremula** Warb. in Engl., Bot. Jahrb. **20**: 171 (1894). —Hutch. in F.T.A. **6**, 2: 137 (1916). —K. Coates Palgrave, Trees Southern Africa: 117 (1977). —C.C. Berg in Kew Bull. **43**: 95 (1988); in F.T.E.A., Moraceae: 77 (1989); in Kirkia **13**: 274 (1990). —van Greuning in S. Afr. J. Bot. **56**, 6: 629, fig. 25 (1990). Type from Tanzania.
    Ficus pulvinata Warb. in Engl., Bot. Jahrb. **20**: 169 (1894). Type from Tanzania.

Tree up to 10 m. tall or a shrub, hemi-epiphytic and strangling or secondarily terrestrial, sometimes a liana. Leafy twigs 1–3(5) mm. thick, sparsely minutely puberulous, periderm not flaking off. Leaves spirally arranged; lamina oblong to elliptic or subobovate to ovate-lanceolate, 2.5–8 × 0.7–3 cm., subcoriaceous to chartaceous; apex subacute to somewhat acuminate; base obtuse to rounded or emarginate; margin entire; both surfaces glabrous, or the midrib puberulous beneath, lateral veins 6–9 pairs; tertiary venation reticulate; petiole 0.7–3(4) cm. long, c. 0.5(1) mm. thick; stipules 2–10 mm. long, up to 3 cm. long on new flush, glabrous, caducous. Figs 1–6 together on curved spurs up to 2 cm. long on the older wood; bud scales of the spurs glabrous or nearly so; peduncle 5–20 mm. long, 0.5–1 mm. thick; basal bracts c. 3 mm. long, free parts caducous or sometimes subpersistent. Receptacle subglobose to ellipsoid, 2–2.5 cm. in diam. when fresh, 1–1.5 cm. in diam. when dry, sparsely or densely minutely puberulous to almost glabrous, ?green at maturity, not or hardly wrinkled and slightly stipitate when dry, wall 0.5–1 mm. thick when dry.

Malawi. S: Zomba Distr., Chikwenga Village, Chingala, 24.vi.1989, *Salubeni & Kaunda* 5430 (MAL). **Mozambique**. MS: Beira, 14.xii.1906, *Johnson* s.n. (K). GI: Gaza, Bilene, 25.ix.1978, *Schäfer* 6538 (K; LMU). M: Inhaca Island, 18.vii.1959, *Mogg* 29433 (K).
Also from East Africa and S. Africa (N. Natal). Coastal and lowland dry evergreen forest and coastal bushland; 0–600 m.
Subsp. *kimuenzensis* (Warb.) C.C. Berg occurs from N. Angola to SE. Nigeria at low altitudes. Subsp. *acuta* (De Wild.) C.C. Berg occurs in W. Kenya, Uganda, Rwanda, Burundi and E. Zaire, in upland rain forest between 1900–2400 m.

30. **Ficus polita** Vahl, Enum. Pl. **2**: 182 (1805). —Hutch. in F.T.A. **6**, 2: 124 (1916); in F.C. **5**, 2: 532 (1925). —Lebrun & Boutique in F.C.B. **1**: 135, t. 14 (1948). —Keay in F.W.T.A. ed. 2, **1**: 611 (1958). —K. Coates Palgrave, Trees Southern Africa: 112 (1977). —C.C. Berg et al. in Fl. Cameroun **28**: 227, t. 81 (1985). —C.C. Berg in Kew Bull. **43**: 93 (1988); in F.T.E.A., Moraceae: 80 (1989); in Kirkia **13**: 275 (1990). Type from Ghana.
    Urostigma politum (Vahl) Miq. in Hook., Lond. Journ. Bot. **6**: 553 (1847).
    Ficus umbrosa Sim, For. Fl. Port. E. Afr.: 100, t. 88 (1909). Type: Mozambique, Maganja da Costa, *Sim* 5549 (K, holotype).

Tree up to 15(40) m. tall, hemi-epiphytic or secondarily terrestrial. Leafy twigs 2–5 mm. thick, glabrous or minutely yellowish puberulous, periderm not flaking off. Leaves spirally arranged; lamina ovate to ± elliptic or oblong, (2)5–16(24) × (1.5)3.5–10(15) cm., subcoriaceous, at least the midrib beneath (and petiole) often drying blackish; apex acuminate; base cordate to truncate or rounded, sometimes to subacute; margin entire; both surfaces glabrous; lateral veins (3)4–12 pairs, tertiary venation partly scalariform to reticulate; petiole 2–12 cm. long, 1–2 mm. thick; stipules 0.5–2 cm. long, glabrous, caducous. Figs 1–4 together on spurs up to 3 cm. long on the older wood; bud scales of the spurs glabrous or nearly so; peduncle 8–20 mm. long; basal bracts 3–4 mm. long, persistent. Receptacle globose to obovoid often shortly stipitate, at least when dry, 2–4 cm. in diam. when fresh, 1.5–4 cm. in diam. when dry, whitish puberulous to greenish-purplish at maturity; wall 2–3 mm. thick and wrinkled when dry.

Subsp. **polita** —C.C. Berg in Kew Bull. **43**: 93 (1988); in Kirkia **13**: 275 (1990). —van Greuning in S. Afr. J. Bot. **56**, 6: 629, fig. 24 (1990).

Lamina ovate to elliptic; lateral veins 5–8(9) pairs. Peduncle 1–2 cm. long. Receptacle (2)3–4 cm. in diam. when fresh, 2–4 cm. in diam. when dry.

Zimbabwe. E: Mutasa Distr., Eastern Highlands Tea Estate, north end, Nyamingura R., c. 880 m., 13.xi.1980, *Pope & Müller* 1681A (SRGH). **Malawi**. S: Malawi Hills Forest, 13.viii.1983, *Dowsett-Lemaire* 926 (FHO). **Mozambique**. N: Memba, c. 500 m., 6.xii.1963, *Torre & Paiva* 9452 (PRE). Z: Quelimane town, 17.x.1963, *Gomes e Sousa* 4810 (PRE). MS: 39 km. NE. of Inhamitanga, 4 km. N. of railway line, 5.xii.1971, *Müller & Pope* 1906 (SRGH). GI: Guijá, 26.vi.1947, *Pedro & Pedrógão* 2095 (PRE). M: E. of Goba, 29.x.1942, *Hornby* 2842 (PRE).
Also known from West Africa, Uganda, Kenya and Tanzania, Angola and S. Africa (Natal), and in Madagascar. Low altitude mixed evergreen forest, mixed high rainfall woodland, riverine and coastal forest; 0–1000 m.

Subsp. **brevipedunculata** C.C. Berg in Kew Bull. **43**: 93 (1988); in F.T.E.A., Moraceae: 80 (1989); in Kirkia **13**: 275 (1990). Type: Malawi, Misuku, Wilindi Forest, *Chapman* 254 (FHO, holotype; K; PRE, isotypes).

Lamina oblong to subovate; lateral veins (8)10–12 pairs. Peduncle 0.8–1.2 cm. long. Receptacle 2–2.5 cm. in diam. when fresh, 1.5–2 cm. in diam. when dry.

Zambia. N: Isoka Distr., Mafinga Mts., c. 2150 m., 21.xi.1952, *Angus* 810A (FHO; K). **Malawi**. N: Misuku, Mapita, 14.iv.1981, *White* 13089 (FHO).
Also in Tanzania. Submontane evergreen rain forest; probably confined to altitudes above c. 1500 m.
This subspecies may be distinguished from subsp. *polita* by its greater number of lateral veins in the leaf blade, its shorter peduncles and probably by its smaller figs.

31. **Ficus chirindensis** C.C. Berg in Kew Bull. **43**: 78, fig. 1 (1988); in F.T.E.A., Moraceae: 79 (1989); in Kirkia **13**: 275 (1990). Type: Zimbabwe, Chipinge Distr., Chirinda Forest, *Goldsmith* 23/64 (SRGH, holotype; BR; K; LISC; WAG, isotypes).

Tree up to 35 m. tall, often with pillar-roots, hemi-epiphytic, secondarily terrestrial. Leafy twigs 1.5–3 mm. thick, minutely puberulous, mostly dark-brown to blackish when dry, periderm not flaking off. Leaves spirally arranged; lamina oblong to elliptic or subovate, (4)6–12(16) × (2.5)3–5.5(7.5) cm., subcoriaceous; apex subacuminate; base cordate to emarginate or truncate rarely rounded or obtuse; margin entire; superior surface glabrous, usually minutely puberulous on the midrib, inferior surface minutely puberulous on the main veins, sometimes glabrous; lateral veins (6)8–12 pairs, tertiary venation reticulate; petiole (1.5)2–4(6) cm. long, 1.5–2 mm. thick; stipules 3–5 mm. long, up to 4 cm. long on new flush, with appressed (on long stipules ± patent) white hairs, caducous. Figs up to 3 together on spurs up to 2 cm. long on the older wood, down to the main branches; bud scales of the spurs glabrous or sparsely puberulous; peduncle (10)15–20(40) mm. long, c. 1.5 mm. thick; basal bracts c. 2.5(3) mm. long, caducous. Receptacle subglobose, 2.5–4 cm. in diam. when fresh, 1.5–3 cm. in diam. when dry, minutely puberulous, greenish to pale-yellow with brown spots at maturity; wall c. 1 mm. thick when dry.

Zimbabwe. E: Mutare Distr., Vumba Mts., 32°44′E, 19°07′S, 11.xi.1982, *Berg, Müller & Campbell* 1355 (K; U). **Malawi**. N: Misuku Hills, Mugesse (Mugeshi) Forest, 27.xi.1983, *Dowsett-Lemaire* 642 (FHO). C: Ntchisi Forest, 1450–1600 m., 30.i.1983, *Dowsett-Lemaire* 587 (FHO; K). S: Thondwe, Mpita (A.L.C.) Tobacco Estate, 1100 m., 4.v.1982, *Chapman & Tawakali* 6167 (K; MAL). **Mozambique**. T: Monte Zóbuè, 1200 m., 3.x.1942, *Mendonça* 578 (LISC).
Also known from Kenya and E. Zaire. Submontane mixed evergreen forest and wooded, dry rocky mountain slopes; 1000–1600 m.

32. **Ficus sansibarica** Warb. in Engl., Bot. Jahrb. **20**: 171 (1894). —Hutch. in F.T.A. **6**, 2: 130 (1916); in F.C. **5**, 2: 532 (1925). —K. Coates Palgrave, Trees Southern Africa: 113 (1977). —C.C. Berg in

Kew Bull. **43**: 93 (1988); in F.T.E.A., Moraceae: 79 (1989); in Kirkia **13**: 276 (1990). Type from Tanzania.

Tree up to 20(40) m. tall, hemi-epiphytic and strangling or secondarily terrestrial. Leafy twigs 2–5 mm. thick, glabrous or sparsely minutely puberulous, periderm of older parts often flaking off when dry. Leaves spirally arranged; lamina oblong to lanceolate, elliptic or ovate, 4.5–13(24) × 2–6(11.5) cm., ± coriaceous, at least the midrib beneath (and the petiole) usually drying red-brown; apex acuminate to subacute or obtuse; base rounded to subcordate or cuneate; margin entire; both surfaces glabrous; lateral veins 5–10(14) pairs; tertiary venation predominantly reticulate; petiole (0.8)2–5.5(8) cm. long, 1–2(3) mm. thick; stipules 1–15 mm. long, up to 4.5 cm. long on new flush, sparsely to densely puberulous or only ciliolate, caducous or subpersistent on new flush. Figs 2–4 together on short, branched and finally ± cushion-shaped spurs up to c. 2(5) cm. long, or on straight, peg-like or sometimes curved spurs up to 15 cm. long, on the main or lesser branches; bud scales of the spurs densely puberulous; peduncle 12–25(50) mm. long, 2–3 mm. thick; basal bracts 3–5 mm. long, free parts caducous or sometimes subpersistent. Receptacle subglobose, often stipitate when dry, 2–6(10) cm. in diam. when fresh, 1.5–3(6) cm. in diam. when dry, puberulous, greenish or partly purplish at maturity; wall 5–10 mm. thick when fresh, 2–4 mm. thick and ± wrinkled when dry.

Subsp. **sansibarica** —C.C. Berg in Kew Bull. **43**: 94 (1988); in F.T.E.A., Moraceae: 79 (1989); in Kirkia **13**: 276 (1990). —van Greuning in S. Afr. J. Bot. **56**, 6: 628, fig. 23 (1990).
   Ficus langenburgii Warb. in Engl., Bot. Jahrb. **30**: 293 (1901). Type from Tanzania.
   Ficus delagoensis Sim, For. Fl. Port. E. Afr.: 99, t. 92 (1909). Type: Mozambique, Maputo (Delagoa Bay), *Sim* 5171 (K, holotype).

Lamina usually oblong to lanceolate, often less than 10 cm. long; lateral veins 5–10(14) pairs; stipules ciliolate; fig-bearing spurs up to 3.5(5) cm. long.

Zimbabwe. N: Gokwe Distr., Sengwa Res. Station, 3.xi.1975, *Guy* 2369 (K; PRE; SRGH). W: Hwange Distr., Kariba–Kamativi road, vii.1954, *Orpen* 46/54 (K; SRGH). E: Mutare Distr., W. base of Murahwa's Hill, 15.ii.1969, *Chase* 8526 (PRE; SRGH). S: Chibi Distr., Nyoni Mts., 28.ii.1970, *Müller & Gordon* 1306 (K; PRE; SRGH). **Malawi**. C: Chitala R., at bridge on Chitala-Salima road, 13.ii.1959, *Robson* 1580 (BM; K; LISC; PRE; SRGH). **Mozambique**. N: Cabo Delgado Distr., ruins on Vamizi (Muamizi) Island, 4.v.1959, *Gomes e Sousa* 4457 (COI; K; PRE). Z: Serra Morrumbala, Massingire, 6.v.1942, *Torre* 4502 (LISC). T: Planalto de Songo, 21.i.1973, *Torre, Carvalho & Ladeira* 18822 (LISC). MS: Gorongosa Nat. Park, Urema R., 8 km. downstream from crossing, ix.1972, *Tinley* 2727 (LISC). GI: Lhanguéne, 10.vii.1947, *Pedro & Pedrógão* 1431 (K; LMA). M: Inhaca Island, Hlangani Hill, 29.ix.1958, *Mogg* 28378B (K; PRE).
   Also known from Kenya, Tanzania and S. Africa (NE. Transvaal), extending westwards up the Zambezi Valley. Low altitude riverine and evergreen forests, miombo and coastal woodlands. Often as a strangler-fig. Cultivated in Harare parks (*Biegel* 2652 (K)).

Subsp. **macrosperma** (Mildbr. & Burret) C.C. Berg in Kew Bull. **43**: 94 (1988); in F.T.E.A., Moraceae: 80 (1989); in Kirkia **13**: 276 (1990). Lectotype from Cameroon, chosen by Berg in Fl. Gabon **26**: 206 (1984).
   Ficus brachylepis Hiern, Cat. Afr. Pl. Welw. **1**, 4: 1011 (1900). —Hutch. in F.T.A. **6**, 2: 126 (1916). —F. White, F.F.N.R.: 32 (1962). Lectotype from Angola, chosen by Berg in Kew Bull. **43**: 94 (1988).
   Ficus macrosperma Mildbr. & Burret in Engl., Bot. Jahrb. **46**: 223 (1911). Type as for *F. sansibarica* subsp. *macrosperma*.
   Ficus ugandensis Hutch. in Bull. Misc. Inf., Kew **1915**: 321 (1915). Type from Uganda.
   Ficus gossweileri Hutch. in Bull. Misc. Inf., Kew **1915**: 321 (1915). Type from Angola.

Leaf lamina elliptic to oblong or ± ovate, often more than 10 cm. long, or leaf oblong to lanceolate and usually less than 10 cm. long; lateral veins 8–12 pairs; stipules glabrous; fig-bearing spurs up to 10(15) cm. long.

Zambia. B: Zambezi Distr., c. 16 km. S. of Zambezi (Balovale) on Chitokoloki road, 11.x.1952, *Angus* 619 (FHO; K). W: Copperbelt Prov., c. 21 km. on Kitwe–Chingola road, 24.xi.1982, *Berg & Bingham* 1413 (K; U). C: 1 km. N. of Kafue bridge, 17.xi.1982, *Berg & Bingham* 1369 (K; U). S: Southern Prov., escarpment, 17.xi.1982, *Berg & Bingham,* 1375 (K; U).
   Also in West tropical Africa, Uganda and Angola. Riverine vegetation, high rainfall miombo, also in *Parinari excelsa* forest on river-bank sand, in muteshi forest, often on Kalahari Sand.

33. **Ficus bubu** Warb. in Ann. Mus. Congo, Bot., sér. 6, **1**: 3, t. 8 (1904). —Hutch. in F.T.A. **6**, 2: 166 (1916). —Lebrun & Boutique in F.C.B. **1**: 160 (1948). —C.C. Berg et al. in Fl. Cameroun **28**: 228,

t. 82 (1985). —C.C. Berg in F.T.E.A., Moraceae: 81 (1989); in Kirkia **13**: 276 (1990). —van
Greuning in S. Afr. J. Bot. **56**, 6: 626, fig. 22 (1990). Type from Zaire.
 Ficus pachypleura Warb. in Ann. Mus. Congo, Bot. sér. 6, **1**: 4 (1904). Lectotype from Zaire,
chosen by Berg in Fl. Gabon **26**: 215 (1984).
 Ficus kyimbilensis Mildbr. in Willldenowia **1**: 27 (1953). Type from Tanzania.

Tree up to 20(30) m. tall, hemi-epiphytic, often terrestrial; bark pale-green to whitish.
Leafy twigs 6–12 mm. thick, glabrous or minutely puberulous, periderm often flaking off
when dry. Leaves spirally arranged; lamina elliptic or sometimes oblong to subcircular,
12–30 × 6–23 cm., coriaceous; apex shortly acuminate to almost rounded; base obtuse to
rounded or (especially in large leaves) cordate; margin entire; both surfaces glabrous;
lateral veins 6–8(9) pairs, often furcate far from the margin; tertiary venation partly
scalariform; petiole 3.5–11(16) cm. long, 2–5 mm. thick; stipules 3–5 mm. long, up to 4 cm.
long on new flush, glabrous or ± puberulous, caducous. Figs on short (often almost
spine-like) spurs on the main branches (or the trunk); bud scales of the spurs minutely
puberulous, mostly apiculate, sublepidote when dry; peduncle 7–10 mm. long, 2–2.5 mm.
thick; basal bracts 4–5 mm. long, persistent. Receptacle globose, c. 3 cm. in diam. when
fresh, c. 2.5 cm. in diam. when dry, glabrous or minutely puberulous, brownish at
maturity; wall wrinkled when dry.

Zimbabwe. E: Chimanimani Distr., W. bank of Haroni R., 1 km. below gorge, 500 m., 10.i.1969,
Biegel 2781 (SRGH). **Malawi.** N: Karonga, xi.1887, *Scott* s.n. (K). C: Grand Beach Hotel, 28.vii.1951,
Chase 3872 (BM; K; SRGH). S: Zomba Distr., Lake Chilwa, 22.ix.1970, *Müller* 1691 (K; SRGH).
Mozambique. N: Nampula, prox. da captaeao de água da Vila, 1.xi.1942, *Mendonça* 1192
(LISC). MS: E. slopes of Mt. Zembe, c. 800 m., 19.vii.1970, *Müller & Gordon* 1329 (K; SRGH).
GI: Guijá Distr., Caniçado, Limpopo R., 15.v.1948, *Torre* 7843 (LISC).
 Also known from the Ivory Coast through to tropical E. Africa extending southwards to Angola
and S. Africa. High rainfall woodland, below evergreen forest, low altitude tree savanna, riverine
vegetation and lake-sides; 0–1200 m.
 This species has a distinctive tall straight trunk with a whitish bark.

34. **Ficus ovata** Vahl, Enum. Pl. **2**: 185 (1805). —Hutch. in F.T.A. **6**, 2: 164 (1916). —Lebrun &
 Boutique in F.C.B. **1**: 160 (1948). —Keay in F.W.T.A. ed. 2, **1**: 608 (1958). —C.C. Berg et al. in Fl.
 Cameroun **28**: 230, t. 83 (1985). —C.C. Berg in F.T.E.A., Moraceae: 81 (1989); in Kirkia **13**: 277
 (1990). Type from Ghana.
  Urostigma ovatum (Vahl) Miq. in Hook., Lond. Journ. Bot. **6**: 553 (1847).
  Ficus buchneri Warb. in Engl., Bot. Jahrb. **20**: 157 (1894). Lectotype from Angola, chosen by
 Berg in Fl. Gabon **26**: 216 (1984).
  Ficus pseudo-elastica Hiern, Cat. Afr. Pl. Welw. **1**, 4: 996 (1900). Type from Angola.
  Ficus tuberculosa var. *elliptica* Hiern, Cat. Afr. Pl. Welw. **1**, 4: 1000 (1900). Type from Angola.
  Ficus octomelifolia Warb. in Ann. Mus. Congo, Bot., sér. 6, **1**: 1 (1904). Type from Zaire.
  Ficus megaphylla Warb. in Ann. Mus. Congo, Bot., sér. 6, **1**: 1 (1904). Type from Zaire.
  Ficus sapinii De Wild. in Fedde, Repert. **12**: 302 (1913). Lectotype from Zaire, chosen by Berg
 in Fl. Gabon **26**: 218 (1984).
  Ficus asymmetrica Hutch. in Bull. Misc. Inf., Kew **1915**: 336 (1915). Type from Angola.
  Ficus brachypoda Hutch. in Bull. Misc. Inf., Kew **1915**: 339 (1915); in F.T.A. **6**, 2: 189 (1916).
 —F. White, F.F.N.R.: 33 (1962) non Miq. (1847). Type from Uganda.

Tree up to 10(25) m. tall, hemi-epiphytic and strangling or terrestrial, sometimes a
shrub or a liane. Leafy twigs 6–12 mm. thick, densely white- or yellowish-puberulous to
pubescent, or almost glabrous, periderm not flaking off. Leaves spirally arranged; lamina
ovate to elliptic, subovate or oblong (5)9–31 × (3.5)6–20 cm., coriaceous; apex acuminate;
base cordate to truncate, obtuse or subacute; margin entire; superior surface glabrous, or
sparsely puberulous in the lower part of the lamina, inferior surface sparsely to densely
white (or brownish) puberulous to hirtellous or glabrous; lateral veins 10–14 pairs, the
basal pair ± faintly branched, ending at the margin below the middle of the lamina;
tertiary venation partly scalariform; petiole 3–10(13) cm. long, 2–4 mm. thick; stipules
3–10 mm. long, glabrous or puberulous to pubescent, caducous. Figs solitary (or in pairs)
in the leaf axils, just below the leaves, or sometimes also on the older wood, initially
enveloped by an ovoid calyptrate bud-cover up to 2.5 cm. long; peduncle up to 5 mm. long,
4–6 mm. thick; basal bracts 3–4 mm. long, persistent. Receptacle ovoid to ellipsoid,
sometimes subglobose or obovoid, 3–5 × 2.5–4.5 cm. in diam. when fresh, 1–4 × 1–3 cm. in
diam. when dry, puberulous to pubescent, greenish at maturity; wall 3–5 mm. thick when
fresh, 1.5–3 mm. thick when dry.

Zambia. N: Mansa Distr., laterite cliff overhanging Lake Bangweulu, 22.viii.1952, *Angus* 257
(FHO; PRE). W: Ndola, growing at edge of Ndola Golf Course, 22.x.1952, *Angus* 646 (FHO;

K). C: Luangwa Game Reserve, escarpment, Mutinondo stream, 23.viii.1966, *Astle* 4968 (K). **Malawi**. N: c. 45 km. N. of Mzimba, 9.vi.1938, *Pole-Evans & Erens* 655 (K; PRE). **Mozambique**. N: Metonia, Massangulo, Junto a Missão Católica, 12.x.1942, *Mendonça* 799 (LISC).
Widespread in subsaharan Africa from Senegal to Ethiopia, and southwards to N. Angola and Mozambique. Swamp forest (mushitu) margins, dambos and lake shores, miombo and escarpment woodlands. Planted as a street tree in Mbala (*Angus* 2874); 0–2100 m.

35. **Ficus cyathistipula** Warb. in Engl., Bot. Jahrb. **20**: 173 (1894). —Hutch. in F.T.A. **6**, 2: 153 (1916). —Lebrun & Boutique in F.C.B. **1**: 172 (1948). —F. White, F.F.N.R.: 32, fig. 6H (1962). —C.C. Berg et al. in Fl. Cameroun **28**: 246, t. 89 (1985). —C.C. Berg in Kew Bull. **43**: 82 (1988); in F.T.E.A., Moraceae: 83 (1989); in Kirkia **13**: 271 (1990). Lectotype from Tanzania, chosen by Berg in Fl. Gabon **26**: 236 (1984).
    Ficus callescens Hiern, Cat. Afr. Pl. Welw. **1**, 4: 1001 (1900). Lectotype from Angola, chosen by Berg in Fl. Gabon **26**: 234 (1984).
    Ficus rhynchocarpa Mildbr. & Burret in Engl., Bot. Jahrb. **46**: 235 (1911). Lectotype from Tanzania, chosen by Berg in Fl. Gabon **26**: 234 (1984).
    Ficus nyanzensis Hutch. in Bull. Misc. Inf., Kew **1915**: 327 (1915). Type from Uganda.

Tree up to 8(15) m. tall, terrestrial or hemi-epiphytic. Leafy twigs 3–5 mm. thick, glabrous or white-puberulous, periderm sometimes flaking off when dry. Leaves spirally arranged; lamina oblanceolate to obovate, 6–20 × 3–7 cm., coriaceous; apex acuminate; base cuneate to attenuate; margin entire; both surfaces glabrous; lateral veins 5–7(8) pairs, tertiary venation reticulate; petiole 1.5–4 cm. long, c. 3 mm. thick, epidermis not flaking off when dry; stipules partly connate, 15–25 mm. long, minutely white-puberulous or almost glabrous, persistent. Figs 1(3) in the leaf axils; peduncle 5–25 mm. long; basal bracts c. 4 mm. long, persistent. Receptacle globose to obovoid (usually on stipes up to 10 mm. long) or pyriform, 3–5 cm. in diam. when fresh, 2–3 cm. in diam. when dry, often somewhat scabrous, pale-green to pale-yellow at maturity; wall up to 8 mm. thick and spongy, 2–5 mm. thick when dry, smooth, verrucose or sometimes with protuberances up to 3 mm. long, apex up to 10 mm. when dry.

Zambia. N: Kundabwika Falls on Kalungwishi River, W. of Mporokoso, 14.iv.1961, *Phipps & Vesey-FitzGerald* 3161 (K; LISC; PRE; SRGH). W: Kabompo R., near Kabompo Boma, 6.x.1950, *Angus* 614 (BM; FHO; K). **Malawi**. N: Nkhata Bay Distr., c. 22 km. E. of Mzuzu, Manchire Falls, 19.vii.1969, *Pawek* 2573 (K).
Also known from the Ivory Coast through to Uganda, Kenya and Tanzania and to N. Angola. Riverine forest or swamp forest (mushitu), often on rocks or river banks or lake shores; 1000–1800 m.
    Ficus cyathistipula subsp. *pringsheimiana* (Braun & K. Schum.) C.C. Berg, with sessile figs, occurs in Cameroon.

36. **Ficus scassellatii** Pamp. in Boll. Soc. Bot. Ital. **1915**: 14 (1915). —Friis in Nordic Journ. Bot. **5**: 331 (1985). —C.C. Berg in Kew Bull. **43**: 94 (1988); in F.T.E.A., Moraceae: 84 (1989); in Kirkia **13**: 271 (1990). Lectotype from Somalia, chosen by Friis loc. cit.
    Ficus kirkii Hutch. in Bull. Misc. Inf., Kew **1915**: 343 (1915); in F.T.A. **6**, 2: 209 (1917). —K. Coates Palgrave, Trees Southern Africa: 110 (1977). Lectotype from Tanzania, chosen by Berg in Kew Bull. **43**: 94 (1988).
    Ficus michelsonii Boutique & J. Léonard in Bull. Jard. Bot. Brux. **19**: 217 (1949). Type from Zaire.

Tree up to 50 m. tall, hemi-epiphytic, secondarily terrestrial. Leafy twigs 3–8 mm. thick, glabrous or minutely puberulous, periderm of older parts flaking off when dry. Leaves spirally arranged; lamina oblong to oblanceolate, obovate or elliptic (6)10–20(28) × 3–8(10) cm., coriaceous; apex shortly and usually bluntly acuminate to rounded; base subacute; margin entire; both surfaces glabrous; lateral veins 8–18 pairs, gradually becoming stronger towards the apex, tertiary venation reticulate; petiole 5–25(35) mm. long, 2–3 mm. thick, epidermis not flaking off when dry; stipules 3–20 mm. long, free, glabrous or minutely puberulous, caducous. Figs solitary or in pairs in the leaf axils; peduncle 5–15 mm. long; basal bracts 3–5 mm. long, persistent. Receptacle globose to ellipsoid, often shortly stipitate at least when dry, 4–4.5 cm. in diam. when fresh, 1.2–2 cm. in diam. when dry, sparsely minutely puberulous, green at maturity; wall (1)4–5 mm. thick, not spongy, apex (in dry material) protruding up to 7 mm.

Zimbabwe. E: Chipinge Distr., Chirinda Forest, vi.1964, *Goldsmith* 21/64 (K; LISC; PRE; SRGH). **Malawi**. N: Chitipa Distr., Misuku Hills, Mugesse Forest Reserve, 15.ix.1970, *Müller* 1657 (K; PRE; SRGH). C: Ntchisi Mt., 20.ii.1959, *Robson & Steele* 1691 (BM; K; SRGH). S: Mulanje Distr., Chiphalombe R., c. 1400 m., 12.viii.1986, *J.D. & C.G. Chapman* s.n. (MAL). **Mozambique**. Z: Milanje Mt., 26.ii.1943, *Torre* 4846 (LISC). MS: Mt. Gorongosa, SE. slopes, 25.vii.1970, *Müller & Gordon* 1460 (LISC; SRGH).

Also from Kenya, Tanzania, Somalia and Zaire. Mixed evergreen forest; 1000–1850 m.

F. scassellatii can be distinguished from the closely related *F. cyathistipula* by the latter having persistent stipules and a spongy fig wall.

Ficus scassellatii subsp. *thikaensis* C.C. Berg occurs in Kenya, and may be distinguished by its leaves up to 28 × 4.5 cm., mostly oblanceolate, its sessile or subsessile figs and its receptacle 2–3 cm. in diam. when dry, apex hardly protruding.

37. **Ficus ardisioides** Warb. in Engl., Bot. Jahrb. **20**: 171 (1894). —Hutch. in F.T.A. **6**, 2: 207 (1917). —F.White, F.F.N.R.: 33 (1962). —C.C. Berg in Kew Bull. **43**: 77 (1988); in Kirkia **13**: 270 (1990). Type from Zaire.

Shrub or tree up to 5(8) m. tall, hemi-epiphytic, sometimes a liane. Leafy twigs 2–3(6) mm. thick, glabrous or sparsely minutely puberulous, periderm not flaking off when dry. Leaves spirally arranged; lamina oblong to elliptic, lanceolate or oblanceolate, (1.5)4–16(26) × (1)1.5–6.5(9) cm., ± coriaceous; apex acuminate to subcaudate; base obtuse; margin entire; both surfaces glabrous; lateral veins (3)4–6(7) pairs, loop-connected 1–5 mm. from the margin, basal pair unbranched, tertiary venation reticulate; petiole 5–15(20) mm. long, 1–2 mm. thick, epidermis usually flaking off when dry; stipules 2–10 mm. long, glabrous, caducous, sometimes subpersistent. Figs in pairs in the leaf axils, sessile (or on peduncles 2–9 mm. long); basal bracts 1.5–2 mm. long, persistent. Receptacle subglobose to obovoid, 1–2 cm. in diam. when fresh, 0.5–1 cm. in diam. when dry, sparsely puberulous, smooth or pusticulate; wall thin; apex plane or (in large figs) prominent when dry.

Subsp. **camptoneura** (Mildbr.) C.C. Berg in Kew Bull. **43**: 77 (1988); in Kirkia **13**: 271 (1990). Type from Cameroon.

    Ficus camptoneura Mildbr. in Engl., Bot. Jahrb. **46**: 233 (1911); in Wiss. Ergebn. Deutsch. Zentr.-Afr. Exped. 1907–1908, **2**: 186 (1911). —Hutch. in F.T.A. **6**, 2: 147 (1916). —Keay in F.W.T.A. ed. 2, **1**: 607 (1958). —C.C. Berg et al. in Fl. Cameroun **28**: 238 (1985). Type as above.

    Ficus camptoneura var. *angustifolia* Mildbr. in Engl., Bot. Jahrb. **46**: 234 (1911); in Wiss. Ergebn. Deutsch. Zentr.-Afr. Exped. 1907–1908, **2**: 187 (1911). Type from Zaire.

    Ficus arcuatonervata De Wild. ex Hutch. in Bull. Misc. Inf., Kew **1915**: 339 (1915). Type as for *F. camptoneura* var. *angustifolia*.

Figs sessile.

Zambia. W: Mwinilunga, Kazera R., c. 12 km. W. of Kakoma, 29.ix.1952, *White* 3411 (BM; FHO; PRE).

Also in SE. Nigeria through to E. Zaire. Riverine and swamp forest understorey; 0–1200 m.

Subsp. *ardisioides* occurs in the Central African Republic and N. Zaire, and is distinguished in having figs on peduncles 2–9 mm. long.

38. **Ficus barteri** Sprague in Gard. Chron., ser. 3, **33**: 354 (1903). —Hutch. in F.T.A. **6**, 2: 205 (1917). —Lebrun & Boutique in F.C.B. **1**: 164 (1948). —Keay in F.W.T.A. ed. 2, **1**: 610 (1958). —F. White, F.F.N.R.: 33 (1962). —C.C. Berg et al. in Fl. Cameroun **28**: 190, t. 65 (1985). —C.C. Berg in F.T.E.A., Moraceae: 83 (1989); in Kirkia **13**: 272 (1990). Type from Nigeria.

    Ficus laurentii Warb. in Ann. Mus. Congo, Bot. sér. 6, **1**: 21 (1904). Type from Zaire.

Tree up to 10 m. tall, hemi-epiphytic, or a shrub. Leafy twigs 3–5 mm. thick, glabrous, periderm not flaking off. Leaves spirally arranged; lamina lanceolate to linear, or less often oblong to elliptic, (5.5)10–18(30) × 1.5–3.5(7) cm., coriaceous; apex acuminate to subacute; base acute to rounded; margin entire; both surfaces glabrous; lateral veins 10–20 pairs, tertiary venation reticulate or parallel to the lateral veins; petiole 1–4.5 cm. long, c. 2 mm. thick, epidermis not flaking off when dry; stipules 5–20 mm. long, glabrous, caducous. Figs in pairs in the leaf axils; peduncle (5)10–25 mm. long; basal bracts 1.5–2 mm. long, caducous. Receptacle globose, 1–1.5 cm. in diam. when fresh, c. 0.5–1 cm. in diam. when dry, glabrous, smooth to verruculose, yellow to orange at maturity.

Zambia. N: Kasama, Chishimba Falls, 26.xi.1952, *Angus* 850 (BM; FHO; K). W: Mufulira, 13.viii.1954, *Fanshawe* 1467 (FHO; K; SRGH).

Also in W. tropical equatorial Africa extending from Sierra Leone to Uganda and southwards to N. Zambia. Riverine and swamp forest; c. 1000–1200 m.

156a. CECROPIACEAE

By C.C. Berg

Trees, often with stilt-roots, or shrubs, dioecious, sap watery turning black. Leaves spirally arranged, lamina palmately or radiately incised, stipulate. Staminate inflorescences branched; tepals 2–4, stamens 1 or 3–4. Pistillate inflorescences globose- or clavate-capitate; tepals 2–3; pistil 1; ovary free or basally adnate to the perianth; stigma 1; ovule 1, basally attached. Fruit achene-like or forming a drupaceous whole with the fleshy perianth. Seed large and without endosperm or small and with endosperm.

A family with 6 genera and c. 200 species, all in the tropics; 1 genus in Asia and Australasia; 3 genera in the Neotropics and 2 genera (9 species) in Africa. *Myrianthus* is recorded for the Flora Zambesiaca area, *Musanga* (with radially incised, peltate leaves) is not.

MYRIANTHUS Beauv.

Myrianthus Beauv., Fl. Oware **1**: 16 (1805). —De Ruiter in Bull. Jard. Bot. Brux. **46**: 472 (1976).
Dicranostachys Tréc. in Ann. Sci. Nat., Bot., Sér. 3, **8**: 85 (1847).
Myrianthus Sect. *Dicranostachys* (Tréc.) Engl. in Mon. Afr. Pflanzen. **1**: 37 (1898).

Trees or shrubs, often with stilt-roots. Lamina basally attached, palmately incised (sometimes entire); stipules fully amplexicaul, connate. Inflorescences bracteate. Staminate inflorescences branched; flowers in spike-like to globose glomerules; tepals 3–4; stamens 3–4. Pistillate inflorescences globose-capitate; perianth 2–3-lobed, stigma tongue-shaped. Fruiting perianth enlarged, fleshy, yellow to orange-red; fruit adnate to the perianth, endocarp woody. Seed large, without endosperm.

The genus is African and comprises 7 species, most of them in West and Central Africa.

Myrianthus holstii Engl. in Mon. Afr. Pflanzen. **1**: 37, t. 16A, 17E (1898). —Rendle in F.T.A. **6**, 2: 237 (1917). —Hauman in F.C.B. **1**: 84 (1948). —F. White, F.F.N.R.: 35 (1962). —De Ruiter in Bull. Jard. Bot. Brux. **46**: 481, t. 4 (1976). —K. Coates Palgrave, Trees Southern Africa: 119 (1977). —C.C. Berg in F.T.E.A., Moraceae: 88 (1989). TAB. **21**. Type from Tanzania.
Myrianthus holstii var. *quinquesectus* Engl. in Bot. Jahrb. **30**: 295 (1901). Type from Tanzania.
Myrianthus mildbraedii Peter in Fedde, Repert., Beih. **40**, 2: 113, Descr.: 10, t. 2, 11 (1932). Syntypes from Tanzania.

Tree up to 20 m. tall. Leaf lamina c. 25 × 25 to 60 × 60 cm., 3–7(8)-fid or parted down to the petiole; segments sessile or stoutly petiolulate, margin regularly serrate-dentate to subentire; superior surface hirtellous to puberulous on the main veins, inferior surface brownish-hirtellous to subsericeous on the main veins; petiole 7–35 cm. long, brownish-hirtellous to subsericeous; stipules 1.5–4 cm. long, brown to golden-yellow sericeous. Staminate inflorescences c. 4–15 cm. in diam.; peduncle 3–13.5 cm. long; glomerules of flowers spike-like, 4–6 mm. in diam. Pistillate inflorescences 1–2 cm. in diam., in fruit 5–8 cm. in diam.; peduncle 1–5 cm. long; flowers c. 20–40; apex of the perianth broadly conical; endocarp body c. 1.2 × 0.8 cm.

Zambia. N: Isoka Distr., Mafinga Hills near Chisenga, 22.xi.1952, *Angus* 822 (FHO; K; PRE). E: Makutus, 27.x.1972, *Fanshawe* 11583 (K; NDO). **Zimbabwe**. E: Chipinge Distr., Chirinda Forest, x.1965, *Goldsmith* 18/65 (K; PRE; SRGH). **Malawi**. N: Chitipa Distr., Misuku Hills, Mugesse Mission, 31.xii.1977, *Pawek* 13487 (K; MAL; MO; PRE). C: Ntchisi Mt., 4.ix.1929, *Burtt Davy* 21055 (FHO). S: Mt. Mulanje, Chisongole Rain Forest, *Newman & Whitmore* 578 (BM; COI). **Mozambique**. N: Ribáuè, Mepáluè, 5.xii.1967, *Torre & Correia* 16374 (LISC). Z: Milange, Serra Tumbine, prox. "Vila Masseti", 18.i.1966, *Correia* 451 (LISC). MS: Serra de Choa, Vila Gouveia, 13.vii.1941, *Torre* 2995 (LISC; LMA; UC).
Also known from Tanzania, Kenya, Uganda and E. Zaire. Submontane evergreen rainforest; 900–2100 m.

Tab. 21. MYRIANTHUS HOLSTII. 1, twig with leaf and staminate inflorescence (× 4), *Goldsmith* 18/65; 2, pistillate inflorescence (× 1), *Wild* 2163. Drawn by Eleanor Catherine.

157. URTICACEAE

By I. Friis

Annual or perennial herbs, shrubs, lianas, or small trees with soft wood; monoecious or dioecious (rarely polygamous); stinging hairs are present in some genera while stiff, non-stinging hairs (without bulbous tip, sack-like base and irritating fluid) are frequent. Stinging hairs are large, unicellular and have a calcified wall, bulbous tip, and a soft-walled, sack-like lower part; the bulbous tip is easily detached, liberating an irritating fluid from the sack-like part; the sack-like base of the stinging hairs is embedded in, or occasionally mounted on, a small epidermal protuberance on stems, petioles or inflorescences, see Tab. 23. Cystoliths (incrustations of calcium or cellulosis most easily visible in dried material) are generally present in the epidermal cells, dot-like, elongated, or linear. Leaves alternate or opposite, petiolate or sessile, sometimes the opposite leaves are anisophyllous with each of a pair of leaves unequal in shape and/or size; stipules usually present, lateral or often intrapetiolar, often fused; lamina simple to deeply 3–5-lobed, rarely 7-lobed, margin entire, serrate or dentate, usually strongly 3-nerved from the base with much smaller lateral nerves from the midnerve, or sometimes evenly penninerved. Inflorescences are extremely varied, mostly cymose, often of densely clustered flowers, and subtended by involucral bracts; occasionally the inflorescence axis is contracted into a flattened, disk-shaped, fleshy receptacle. Flowers minute, unisexual, actinomorphic or (especially the female flowers) zygomorphic, with one whorl of tepals or rarely the female flowers naked, sessile or pedicellate, pedicel often articulated just below the perianth. Male flowers with (1)2–5 tepals; stamens equal in number and opposite the tepals or solitary, inflexed in bud until reflexing suddenly to forcibly eject the pollen; anthers 2-thecous opening by longitudinal slits; rudimentary ovary usually present. Female flowers with 3–5 tepals, which are free or united, often very unequal, often accrescent after pollination, rarely absent; staminodes, when present, scale-like, inflexed (sometimes actively ejecting the achene when ripe); ovary superior, syncarpous, erect, usually somewhat laterally compressed, symmetrical or asymmetrical, glabrous, 1-celled, with 1 erect basal ovule (placenta absent or very small); style usually absent or very short; stigma capitate, brush-like, or linear. Fruit an achene, consisting of the hardened ovary-wall, often enclosed by the persistent accrescent perianth. Seed with a thin membranaceous testa, usually not fused to the endocarp, mostly with little or no endosperm, and the cotyledons correspondingly swollen and fleshy.

A family of about 50 genera comprising some 1000 species; almost cosmopolitan, but most numerous in the tropics. Most species occur in humid habitats, e.g. forest floor and riverine vegetation. Some genera are adapted to dry or extremely dry environments; this applies to *Obetia* and some species of *Pouzolzia* in the Flora Zambesiaca area.

The genera of this family can be difficult to define and identify, and their determination frequently requires close examination of often very small floral details. Confusion with some genera of the *Euphorbiaceae*, especially the genus *Acalypha*, has been frequent in herbaria and in the literature, but the long male inflorescence spike of most species of *Acalypha* is unknown in African *Urticaceae*.

The species of some genera (especially *Obetia*, *Girardinia*, *Laportea*, *Pilea* and *Elatostema*) are extremely plastic, and their identification can be difficult when many species occur in the same area. Weddell, in his monographic treatment of the family (Monogr. Urtic.: 4 (1856)), maintains that in no other plant family are the species and genera so difficult to recognise on habit alone; he further states that within a single species all parts of a plant can be subject to considerable variation. Delimitation and identification of species in the *Urticaceae* can present serious problems.

Key to the genera

1. Plants with stinging hairs (i.e. hairs with a sack-like base, calcified wall and bulbous tip, liberating an irritating fluid), at least on the inflorescences and petioles - - - - - 2
- Plants completely without stinging hairs - - - - - - - - - 5
2. Stipules free, lateral; shrubs or small trees with soft wood - - - - **2. Obetia**
- Stipules fused, intrapetiolar; woody climbers or erect, short-lived or perennial herbs - - - - - - - - - - - - - - - - 3

3. Woody climbers with adventitious or axillary roots from the climbing stems; stinging hairs mostly restricted to petioles and inflorescences; female perianth with tepals accrescent, becoming fleshy and orange or red in fruit　-　-　-　-　-　-　-　-　-　-　**1. Urera**
– Erect, short-lived or perennial herbs, not climbing and without adventitious or axillary roots from the stems; usually with stinging hairs on all aerial parts of the plant; female perianth membranaceous in fruit　-　-　-　-　-　-　-　-　-　-　-　-　4
4. Female perianth of 3 more or less united tepals (sometimes also with one minute free tepal), forming a 1-sided cover on the ovary or achene; stinging hairs often more than 5 mm. long　-　-　-　-　-　-　-　-　-　-　-　-　-　-　-　**3. Girardinia**
– Female perianth of 4 free, very unequal tepals; stinging hairs shorter than 5 mm. in length　-　-　-　-　-　-　-　-　-　-　-　-　-　-　**4. Laportea**
5. Leaves opposite, not or only slightly heteromorphic, the 2 leaves of a pair equal or subequal in size and shape　-　-　-　-　-　-　-　-　-　-　-　-　-　-　6
– Leaves alternate, or if opposite then the 2 leaves of a pair strongly heteromorphic, being unequal in size and/or shape　-　-　-　-　-　-　-　-　-　-　-　-　9
6. Inflorescence a pedunculate capitulum with a disk- or bell-shaped receptacle　　**6. Lecanthus**
– Inflorescence not as above, or if receptacle disk- or bell-shaped then capitulum sessile　　7
7. Stipules fused, intrapetiolar　-　-　-　-　-　-　-　-　-　-　-　**5. Pilea**
– Stipules free, lateral　-　-　-　-　-　-　-　-　-　-　-　-　-　8
8. Male flowers with 4 stamens; inflorescences of long, pendent interrupted spikes　-　-　-　-　-　-　-　-　-　-　-　-　-　**9. Boehmeria**
– Male flowers with only 1 stamen; inflorescences dense axillary clusters　　**11. Droguetia**
9. Leaves distichous or opposite with the 2 leaves of a pair strongly heteromorphic, sessile or subsessile, asymmetrical about the midrib　-　-　-　-　-　-　-　-　-　10
– Leaves spirally arranged, petiolate, symmetrical (or almost so) about the midrib　　11
10. Inflorescences consisting of sessile capitula with ± disk-shaped or lobed receptacles surrounded by involucral bracts; leaves always distichous　-　-　-　-　-　-　**8. Elatostema**
– Inflorescences consisting of pedunculate glomerules or of a pedunculate globular receptacle not surrounded by involucral bracts; leaves distichous or opposite in strongly heteromorphic pairs　-　-　-　-　-　-　-　-　-　-　-　-　-　**7. Procris**
11. Inflorescence a lax panicle or an interrupted spike　-　-　-　-　-　-　12
– Inflorescence an axillary, sessile, ± densely capitulate cluster　-　-　-　-　13
12. Woody climber; female perianth with fused accrescent tepals which increase in size and become orange or red in fruit　-　-　-　-　-　-　-　-　-　-　-　**1. Urera**
– Erect or prostrate, herbaceous or slightly woody plants; female perianth with free tepals, membranous in fruit　-　-　-　-　-　-　-　-　-　-　-　**4. Laportea**
13. Male flowers with 1 stamen; female flowers without a perianth; a minute, annual herb　-　-　-　-　-　-　-　-　-　-　-　-　-　**12. Didymodoxa**
– Male flowers with 4–5 stamens; female flower with a tubular perianth; annual or perennial herbs or shrubs　-　-　-　-　-　-　-　-　-　-　-　**10. Pouzolzia**

1. URERA Gaudich.

Urera Gaudich. in Freyc., Voy. Monde, Bot.: 496 (1830). —Friis in Nordic Journ. Bot. **5**, 6: 547–553 (1986).

Dioecious woody climbers, or lianas often reaching the top of the supporting tree, mostly fixed to the substrate by adventitious, axillary roots from the stems; stinging hairs on the herbaceous parts or at least a few on the petioles and inflorescences, rarely totally absent, frequently mounted on epidermal protuberances. Cystoliths dot-like. Leaves alternate, petiolate; lamina simple, with an entire to dentate margin, triplinerved. Stipules fused for half to two thirds of their length, intrapetiolar. Inflorescences axillary, paniculate cymes. Male flowers 4–5-merous, stamens equalling the tepals in number; ovary rudimentary. Female flowers with 4 somewhat unequal tepals ± free or fused for at least three quarters of their length; staminodes absent; ovary ovoid; stigma almost sessile, penicillate. Fruit an achene, enclosed by the persistent perianth which increases in size and becomes fleshy and orange-yellow or red in fruit.

A genus of about 40 species, widespread in tropical Africa, Madagascar, tropical America and the Pacific Islands. Three distinct species in eastern and S. Africa; more species, of critical delimitation, in W. Africa.

1. Leaf margin entire, never crenulate or dentate; lamina with 2(3) pairs of lateral nerves above the basal pair; raised protuberances with stinging hairs rare, if present normally only on the inflorescences; female perianth almost entirely enclosing the ovary and the young fruit - - - - - - - - - - - - - - - - 1. *trinervis*
- Leaf margin crenulate to serrate or dentate, at least in the upper half; lamina with 3–4(5) pairs of lateral nerves above the basal pair; raised protuberances with stinging hairs present or absent; female perianth enclosing the ovary and young fruit or apparently deeply 2-lobed - - - - - - - - - - - - - - - - 2
2. Leaf margins serrate or dentate; stems and petioles with ± forked protuberances up to 3 mm. high and crowned with stinging hairs; female perianth tubular, almost entirely enclosing the ovary and young fruit, obscurely 4-dentate at the apex - - - - 2. *sansibarica*
- Leaf margins undulate or crenulate (not serrate or dentate); stems and petioles without protuberances, and usually devoid of stinging hairs; female perianth covering about three quarters of the ovary, with 2 minute outer lobes and 2 larger, almost circular inner lobes - - - - - - - - - - - - - - 3. *hypselodendron*

1. **Urera trinervis** (Hochst.) Friis & Immelman in Nordic Journ. Bot. **7**: 126 (1987) —Friis in F.T.E.A., Urticaceae: 6 (1989). TAB. **22**, fig. B1–B3. Type from S. Africa.
   Elatostema trinerve Hochst. in Flora **28**: 88 (1845). Type as above.
   Urera cameroonensis Wedd. in DC., Prodr. **16**, 1: 97 (1869). —Rendle in F.T.A. **6**, 2: 261 (1917). —Hauman in F.C.B. **1**: 185 (1948). —Keay in F.W.T.A., ed. 2, **1**: 618 (1958). —Letouzey in Fl. Cameroun **8**: 76 (1968). —Agnew in Upland Kenya Wild Fl.: 321 (1974). —Friis in Nordic Journ. Bot. **5**, 6: 549, tab. 1, fig. H–K (1986). Type from Cameroon.
   Urera arborea De Wild. & T. Dur. in Bull. Soc. Roy. Bot. Belg. **38**, 2: 52 (1899). Type from Zaire.
   Urera laurentii De Wild., Miss. Laurent.: 72, t. 20 (1905). Type from Zaire.
   Urera gilletii De Wild. in Ann. Mus. Congo, Bot. Sér. 5, **1**: 240 (1906). Type from Zaire.
   Urera woodii N.E. Br., in Bull. Misc. Inf., Kew **1911**: 96 (1911); in F.C. **5**, 2: 549 (1925). Type from South Africa (Natal).
   Urera usambarensis Rendle in Journ. Bot. **54**: 370 (1916); in F.T.A. **6**, 2: 263 (1917). Type from Tanzania.
   Urera cameroonensis var. *laurentii* (De Wild.) Rendle in F.T.A. **6**, 2: 262 (1917). Type as above.
   Urera acuminata var. *cameroonensis* (Wedd.) J. Léandri in Fl. Madag. **56**, Urtic.: 24 (1965) excl. spec. cit. Type as above.

Liana, climbing to 10 m. or more. Stems softly woody, up to 10 cm. in diam. at base; bark grey to brownish-black, longitudinally striate, glabrous, rarely pubescent, with large leaf scars on young branches but very rarely with raised protuberances; sap copious, clear; pith wide, spongy, or stems hollow in centre. Leaves (4)6–12 × 3.5–8 cm., elliptic to ovate, rarely obovate; apex acuminate to caudate; base cuneate, truncate or rounded, rarely subcordate; margin entire; lamina coriaceous, upper surface glabrous, rarely with a few scattered stiff hairs, numerous elongated cystoliths present, lower surface glabrous, occasionally pubescent or with scattered, stiff and perhaps stinging hairs, especially on the lateral nerves, rarely with stinging hairs on raised protuberances up to 0.8 mm. high; lateral nerves 2–3(4) pairs, the basal pair extending into upper third of lamina, all interconnected by scalariform-tertiary nerves. Stipules brown, fused almost to the apex, 6–10 mm. long, pubescent on the midnerves to subsericeous outside, glabrous inside. Petioles 2.5–6 cm. long, glabrous to pubescent, rarely with a few stinging hairs, which are sessile or raised on protuberances up to 0.5 mm. high. Male inflorescences lax, paniculate, c. 6.5 cm. long, with flowers densely clustered at intervals along the axes; peduncle mostly with a few stinging hairs; female inflorescence lax, paniculate, c. 2 cm. long, with a small cymose cluster of flowers, peduncle with stinging hairs, especially around the cluster of flowers. Male flowers on pedicels up to 1 mm. long, 4-merous, glabrous, perianth 1–1.75 mm. in diam. Female flowers sessile; perianth 1–1.5 mm. long, cylindrical, constricted at apex, with 4 blunt but clearly marked teeth, glabrous. Achene glabrous, 1.5–2 mm. long, without markings, enclosed in the persistent, accrescent, fleshy, orange perianth.

Zambia. N: Kundabwika Falls, female fl. 7.x.1958, *Fanshawe* 4872 (K; SRGH). **Zimbabwe**. E: Chirinda, male fl. 19.x.1947, *Wild* 2043 (SRGH). **Mozambique**. Z: Milange, Serra Chiperone, Marrega, 1260 m., female fl., fr. 3.ii.1972, *Correia & Marques* 2499 (LMU). MS: Espungabera Mt. (Spungabera), male fl. 11.x.1943, *Torre* 6155 (LISC).
   Also throughout Central Africa, west to Ghana, east to Ethiopia, south to S. Africa (Natal). A climber in lowland and transitional rain forest, especially at forest edges and in clearings; 300–1300 m.

2. **Urera sansibarica** Engl., Pflanzenw. Ost-Afr. **C**: 162 (1895). —Rendle in F.T.A. **6**, 2: 263 (1917). —Friis in Nordic Journ. Bot. **5**: 549, tab. 1, fig. L–M (1986); in F.T.E.A., Urticaceae: 7 (1989). TAB. **22**, fig. C1–C3. Type from Tanzania.

Urera fischeri Engl., Pflanzenw. Ost-Afr. **C**: 162 (1895). —Rendle in F.T.A. **6**, 2: 264 (1917). Type from Tanzania, destroyed.

Urera braunii Engl., Pflanzenw. Afr. **3**, 1: 53 (1915). Type from Tanzania, not specified.

Liana, climbing to 3 m. or higher. Stems with reddish-brown, sometimes peeling bark, usually (especially on younger parts) with numerous fleshy protuberances up to 5 mm. long, crowned by a number of stinging hairs; young stems densely pubescent to hirtellous, older stems glabrescent. Leaves herbaceous, dark green, drying almost black, with a dense, whitish indumentum on the underside, 7.5–15 × 5–10.5 cm., ovate to elliptic, rarely obovate; apex acuminate; base cordate to subcordate; margin dentate to serrate; upper surface of lamina subscabrous, glabrous or with a few short, stiff hairs, cystoliths dot-like, lower surface with stinging hairs on the nerves and usually densely white-felted between the nerves, triplinerved, with the basal pair of nerves extending to the upper half to one third, midrib with 3–4 pairs of nerves above the basal pair. Petiole 2.5–11.5 cm. long, usually densely covered with raised protuberances with stinging hairs. Stipules 0.5–1.0 cm. long, fused for at least two thirds. Male inflorescences up to 10 cm. in diam., irregularly branched; peduncles 1–2 cm. long; flowers in dense clusters, pedicels c. 1 mm. long, perianth c. 1 mm. in diam., 5-merous. Female inflorescences sessile or on peduncles up to c. 2 cm. long, up to 5 cm. in diam., with short rather fleshy branches, densely covered with stinging hairs; flowers sessile, in dense clusters, each surrounded by clusters of stinging hairs, perianth tubular, 4-merous, ovary enclosed except for a penicillate stigma. Achene compressed, c. 1.5 mm. long, surrounded by the orange, accrescent perianth.

Mozambique. Z: Maganja da Costa, Gobene, 20 m., male fl. 14.ii.1966, *Torre & Correia* 14610 (LISC). MS: Cheringoma to Inhaminga, female fl. 19.xi.1942, *Mendonça* 1464 (LISC).

Also in the coastal area of Kenya and Tanzania. Climbing over rocks and trees in coastal forest and evergreen scrub, apparently often associated with limestone areas.

3. **Urera hypselodendron** (A. Rich.) Wedd. in Ann. Sci. Nat., Bot. Sér. 3, **18**: 203 (1852). —Rendle in F.T.A. **6**, 2: 255 (1917). —Hauman in F.C.B. **1**: 181 (1948). —Agnew in Upland Kenya Wild Fl.: 321 (1974). —Friis in Nordic Journ. Bot. **5**: 551, tab. 1, fig. A–G (1986); in F.T.E.A., Urticaceae: 9 (1989). TAB. **22**, fig. A1–A7. Type from Ethiopia.

Urtica hypselodendron A. Rich., Tent. Fl. Abyss. **2**: 260 (1850). Type as above.

Urera schimperi Wedd., Monogr. Urtic.: 158 (1856) nom. illeg. Based on same type as above.

A woody liana, climbing to 25 m. or more high. Stems with numerous adventitious roots; bark of young stems reddish-brown, pubescent to glabrescent without protuberances and stinging hairs but with numerous large leaf scars; older stems with glabrous, smooth or striate, reddish-brown bark, usually not peeling; sap copious, clear. Leaves 10–18 × 7–13 cm., ovate to obovate; apex short acuminate; base rounded to subcordate; margin finely serrate or crenulate, with more than 35 crenulations on each side; lamina dark green, drying brownish black, upper surface glabrous, occasionally with a few stiff hairs, cystoliths dot-like to elongated, lower surface glabrescent to puberulous especially on the nerves, rarely entirely pubescent, or with a few stinging hairs on the nerves; triplinerved with 3–4(5) pairs of lateral nerves, the basal pair extending into upper third of lamina, tertiary nerves and nerves of higher order clearly marked, scalariform. Stipules dark brown, triangular to lanceolate, fused for at least two thirds of their length, up to c. 3 mm. long, puberulous, often ciliate. Petioles 4–10 cm. long, mostly with a few stinging hairs. Inflorescences elongated, lax cymes 4–18 cm. long, usually with numerous stinging hairs. Male inflorescences sessile or on peduncles up to c. 1 cm. long, usually larger than the female; flowers on c. 2 mm. long pedicels, 4-merous, perianth c. 1.5 mm. in diam. Female inflorescences sessile or on peduncles up to c. 2 cm. long; flowers sessile to subsessile, surrounded by stinging hairs, densely clustered, on pedicels c. 1 mm. long, perianth segments 4, basally fused, very unequal, 2 outer short, 2 inner subcircular and almost as long as the ovary, stigma penicillate, protruding. Achene 1–1.5 mm. long, compressed, slightly oblique, brown, minutely granulate, enclosed by the 2 orange-red, fleshy, accrescent perianth lobes.

Zambia. E: Nyika Plateau, 2150 m., female fl. 25.x.1958, *Robson* 351 (K; SRGH). **Zimbabwe**. E: Nyanga Distr., Kukwanissa, Honzo Mt., female fl. & fr. 5.i.1968, *Chase* 8474 (PRE; SRGH). **Malawi**. N: Rumphi Distr., Nyika Plateau, Kafwimba forest, 1800 m., female fl. & fr. 17.x.1975, *Pawek* 10263 (K; MO; PRE; SRGH). C: Dedza Mt., female fl. 29.x.1965, *Banda* 720 (SRGH). S: Zomba Plateau, female fl. 20.xi.1979, *Banda & Salubeni* 1618 (SRGH). **Mozambique**. Z: Milange, Serra Chiperone, 1450 m., ster. 1.ii.1972, *Correia & Marques* 2462 (LMU). MS: Manica, Serra Zuira, 1900 m., male fl. 9.xi.1965, *Torre & Pereira* 12785 (LISC).

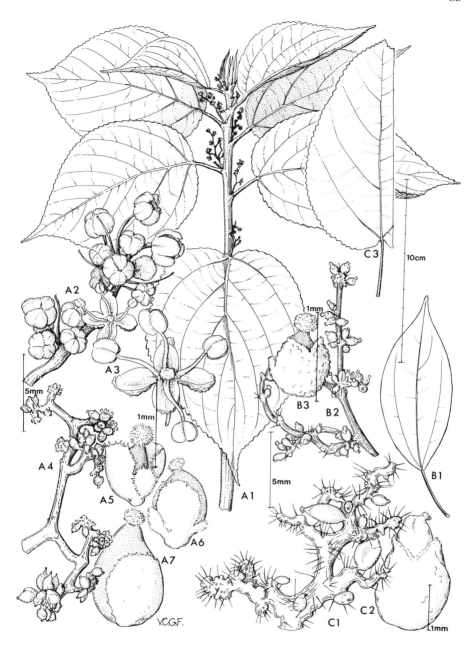

Tab. 22. A.—URERA HYPSELODENDRON. A1, flowering stem, *Chiovenda* 2593; A2, detail of male inflorescence; A3, male flower, A2–3 *Wild* 3570; A4, detail of female inflorescence; A5, young female flower; A6, old female flower; A7, fruit with persisting fleshy perianth, A4–7 *Mooney* 8720. B.—URERA TRINERVIS. B1, leaf, *Hansen* 120; B2, detail of female inflorescence; B3, old female flower with persisting fleshy perianth, B2–3 *Faulkner* 4199. C.—URERA SANSIBARICA. C1, detail of female inflorescence; C2, old female flower with persisting fleshy perianth, C1–2 *Musyoki & Hansen* 954; C3, leaf, *Torre & Correia* 14610. Drawn by Victoria C. Friis. From F.T.E.A.

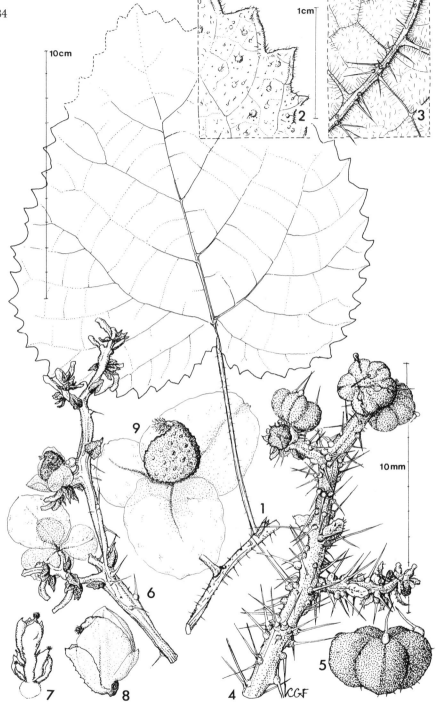

Tab. 23. OBETIA TENAX. 1, young stem with leaf; 2, leaf, superior surface and margin; 3, leaf, inferior surface, 1–3 *Chase* 7942; 4, part of male inflorescence; 5, male flower (in bud), 4–5 *Brain* 11075; 6, part of female inflorescence; 7, young female flower; 8, older female flower with accrescent membranaceous perianth; 9, fruit (achene) with persisting perianth, 6–9 *Chase* 7867. Drawn by Victoria C. Friis.

Also in Kenya, Uganda, Tanzania, E. Zaire, S. Sudan and Ethiopia. A common climber in upland and montane forest clearings, along forest edges, in riverine or ravine forest, or sometimes on isolated trees left in upland farmland; 1450–2150 m.

2. OBETIA Gaudich.

Obetia Gaudich. in Voy. Monde Bonite, Bot., Atlas: tab. 82 (1844). —Friis in Kew Bull. **38**: 221–228 (1983).

Small to medium-sized shrubs or small trees, dioecious, deciduous, wood soft, juicy; younger branches herbaceous, with stinging hairs. Cystoliths dot-like or slightly elongated. Leaves alternate, usually clustered in terminal rosettes, petiolate; lamina simple or lobed, base rounded or cordate, mostly with a velvety tomentum on the inferior side. Stipules free, lateral, mostly persisting. Inflorescences mostly in the axils of fallen leaves, appearing before the new leaves, sometimes in the axils of the new leaves; flowers pedicellate, clustered in branched, bracteate panicles; male inflorescences smaller than the female ones. Male flowers regular, 5-merous, stamens equalling tepals in number, ovary rudimentary. Female flowers irregular, 4-merous, outer pair of tepals smaller than inner pair; tepals increasing in size and becoming thinly membranaceous in fruit; staminodes absent; ovary ovoid, with a sessile penicillate stigma. Achene compressed, enclosed in the persistent accrescent, membranaceous perianth.

A genus of 8 species in E. and S. Africa, Madagascar, the Mascarenes and Aldabra Island. Most species occur in dry habitats.

Obetia tenax (N.E. Br.) Friis in Kew Bull. **38**: 226 (1983). TAB. **23**. Type from S. Africa (Natal).
   Urera tenax N.E. Br. in Hook., Ic. Pl. **18**: t. 1748 (1888); in Bull. Misc. Inf., Kew **1888**: 84 cum tab. (1888); in F.C. **5**, 2: 548 (1925). —Palmer & Pitman, Trees of S. Afr., ed. 2, **1**: 485 (1972). —K. Coates Palgrave, Trees Southern Africa: 121 (1977). Type as above.

A small tree, branched, deciduous, presumably dioecious. Bark of older branches brown. Numerous stinging hairs on younger branches, leaves and inflorescences. Leaves 5–15 × 3–10 cm., ovate to almost circular, sometimes faintly 3-lobed; apex rounded, acute or broadly acuminate; base broadly cordate, truncate or rounded; margin coarsely serrate; lamina triplinerved, upper surface with stinging hairs and a few stiff hairs, lower surface pubescent when young, later glabrescent. Petiole 3–8 cm. long, with stinging hairs. Stipules up to 8 mm. long, lateral, free. Inflorescences in the axis of fallen leaves, developing before or partly contemporaneous with the new leaves. Male inflorescences up to 15 cm. long, bracteate; flowers pedicellate, 5-merous, c. 1 mm. in diam. Female inflorescences as for the males; flowers 4-merous, up to 1 mm. long. Achenes about 1.5 mm. long, enclosed by the accrescent, membranaceous perianth up to 3 mm. long.

   Botswana. SE: Gaborone, 1130 m., female fl. 25.viii.1978, *O.J. Hansen* 3444 (BM; C; GAB; K; PRE; SRGH). **Zimbabwe**. N: Trelawney, 1350 m., male fl. (?a few female ones) i.1941, *Brain* 11075 (SRGH). W: Matobo Distr., Wawa Hill, 1350 m., ster. 8.v.1966, *Best* 483 (SRGH). E: Mutare Distr., Murahwa's (Murahuva's) Hill, W. Commonage, female fl. 14.x.1962, *Chase* 7867 (SRGH). S: Great Zimbabwe, on the Acropolis, 1200 m., male fl. 4.x.1949, *Wild* 3027 (SRGH). **Mozambique**. M: Maputo, between Santos and Frazar, female fl. 2.ix.1948, *Gomes e Sousa* 3820 (COI; PRE).
   Also in S. Africa (Transvaal, Natal, E. Cape Province). In rocky ravines in deciduous bushland, very frequently on granite outcrops; 1000–1400 m.

   Closely related to *O. carruthersiana* (Hiern) Rendle from Namibia and W. Angola, and perhaps no more than subspecifically distinct. *O. carruthersiana* may be distinguished by its cordate leaves and stipules up to c. 12 mm. long.

3. GIRARDINIA Gaudich.

Girardinia Gaudich. in Freyc., Voy. Monde, Bot.: 498 (1830). —Friis in Kew Bull. **36**: 143–157 (1981).

Erect annual (or short-lived perennial) herbs, monoecious or dioecious by abortion, with long stinging hairs on all aerial parts. Cystoliths dot-like. Leaves alternate, petiolate; lamina elliptic to ovate, mostly variously divided, margin coarsely serrate, always triplinerved. Stipules intrapetiolar, fused almost to the apex. Inflorescences unisexual, consisting of dense, elongate cymes in the axils of upper leaves, male inflorescence thinner and more spike-like than the females. Male flowers pedicellate, 4–5-merous with a rudimentary ovary. Female flowers sessile, with 3 almost completely fused tepals ± enclosing the ovary, sometimes also with one minute free tepal; staminodes absent; ovary ± asymmetrical, reflexed, laterally compressed, with a sessile, filiform stigma. Achene laterally compressed, rugose, released from the perianth.

A genus of 2 species distributed in the mountains of the Old World tropics, from West Africa to SE. China.

This genus has probably the largest stinging hairs in the family, but the effect of the sting is less severe than in other genera, e.g. *Laportea*.

Girardinia diversifolia (Link) Friis in Kew Bull. **36**: 145 (1981); in F.T.E.A., Urticaceae: 13, fig. 4 (1989). TAB. **24**. Type from N. India.
    Urtica diversifolia Link, Enum. Pl. Hort. Berol., Alt. **2**: 385 (1822), non Blume (1825). Type as above.
    Urtica palmata Forssk., Fl. Aegypt.-Arab.: 159 (1775), non *Girardinia palmata* Blume (1855). Type from the Yemen.
    Urtica heterophylla Vahl, Symb. Bot. **1**: 76 (1790) nom. illegit. superfl. Type as for *U. palmata* Forssk.
    Girardinia heterophylla Decne. in Jacquem., Voy. Inde 4, Bot.: 152 (1844). —Wedd., Monogr. Urt.: 164 (1856); in DC., Prodr. **16**, 1: 100 (1869). —Letouzey in Fl. Cameroun **8**: 110, t. 17 (1968). —Wickens in Fl. Jebel Marra: 120 (1976). Type as for *Urtica palmata* Forssk.
    Urtica adoensis Steud. in Flora **33**: 259 (1850). —A. Rich., Tent. Fl. Abyss. **2**: 262 (1851). Type from Ethiopia.
    Urtica condensata Steud. in Flora **33**: 260 (1850). —A. Rich., Tent. Fl. Abyss. **2**: 263 (1851). Type from Ethiopia.
    Girardinia adoensis (Steud.) Wedd. in Ann. Sci. Nat., Bot. Sér. 4, **1**: 181 (1854). Type as above.
    Girardinia condensata (Steud.) Wedd. in Ann. Sci. Nat., Bot. Sér. 4, **1**: 181 (1854); in Monogr. Urtic.: 169, t. 2, fig. 1–5 (1856); in DC., Prodr. **16**, 1: 103 (1869). —Rendle in F.T.A. **6**, 2: 266 (1917). —Hauman in F.C.B. **1**: 196 (1948). —Robyns, Fl. Parc Nat. Alb. **1**: 76 (1948). —F.W. Andr., Fl. Pl. Anglo-Egypt. Sudan **2**: 278 (1952). —Keay in F.W.T.A., ed. 2, **1**: 618 (1958). Type as above.
    Girardinia condensata var. *adoensis* (Steud.) De Wild. in Ann. Mus. Congo Bot., Sér. IV, **1**: 173 (1903). Type as for *Urtica adoensis* Steud.
    Girardinia heterophylla subsp. *adoensis* (Steud.) Cufod. in Bull. Jard. Bot. Nat. Belg. **39**, suppl.: XX (1969). Type as above.

Erect annual or short-lived perennial herbs to 1.5(2) m. tall, monoecious or dioecious by abortion. Stem sparsely branched, pubescent and covered with stinging hairs 7–9 mm. long, and with short stiff hairs. Leaves usually fallen from the lower part of the stem at anthesis, 10–20(25) × 10–18(23) cm., ovate to cordate but extremely variously lobed or divided (lamina of the younger leaves usually the least divided); apices of entire leaves and of lobes acuminate; base cuneate, truncate or cordate; margin dentate (with 20–25 teeth on each side in undivided leaves), teeth fine to coarse; lamina not bullate, triplinerved, upper surface glabrescent except for the stinging hairs, lower surface pubescent and with stinging hairs on the nerves. Petioles 3–15 cm. long, densely beset with stinging hairs. Stipules linear-lanceolate, fused for at least four fifths of their length, usually fallen at anthesis. Cystoliths dot-like. Inflorescences unisexual, cylindrical, up to 10 cm. long. Male inflorescences narrow, spicate panicles on peduncles to c. 2 cm. long; flowers on pedicels c. 1 mm. long, clustered, perianth 4–5-merous, tepals without dorsal appendages, rudimentary ovary present. Female inflorescences thicker than the males, densely cymose, consisting of small dichasia 2–3 cm. long at anthesis but elongating to 10–15 cm. during ripening of fruit; flowers sessile, perianth c. 2 mm. long with 3 fused tepals, a free fourth tepal usually absent, ovary enclosed in perianth, stigma filiform. Achene up to 2 mm. long, ovoid to subcordate, compressed, rugose.

Zambia. N: Lake Mweru, Chiengi, 1000 m., female fl. & fr. 17.vii.1957, *Whellan* 1396 (PRE; SRGH). E: Chipata (Fort Jameson), female fl. & fr. 31.v.1958, *Fanshawe* 4483 (SRGH). **Zimbabwe**. N: Makonde Distr., Mhangura (Mangula), female fl. & fr. 20.iv.1962, *Jacobsen* 1689 (PRE; SRGH). W: Matobo Distr., Besna Kobila, 1600 m., male & female fls. iv.1953, *Miller* 1729 (SRGH). C: Marondera (Marandellas), male & female fls. 6.iv.1950, *Wild* 3319 (M; SRGH). E: Chimanimani (Melsetter) Distr., Gwendingwe Forest, 1590 m., female fl. & fr. 28.vi.1976, *Müller* 2897 (SRGH). S: Kyle Dam, Beza Spring, female fl. & fr. 25.v.1971, *Mavi* 1249A (SRGH). **Malawi**. N: Mzimba Distr., Viphya Plateau, Mzuzu, female fl. 15.v.1976, *Pawek* 11269 (K; MO; SRGH). C: Lilongwe Distr., Dzalanyama, Kawai Hills, female fl. & fr. 29.iv.1958, *Jackson* 2226 (SRGH). S: Thyolo

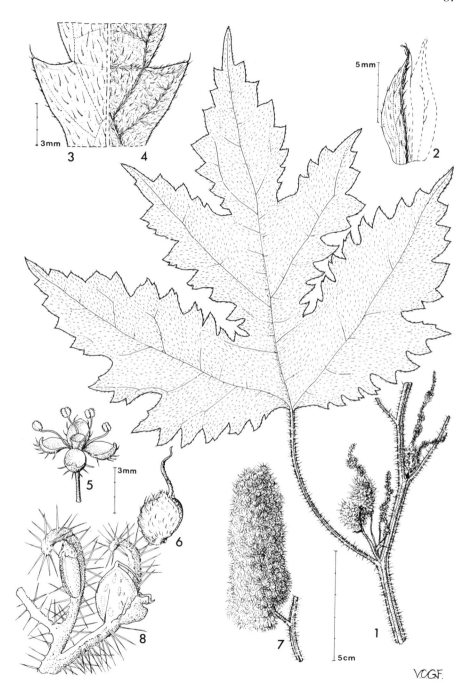

Tab. 24. GIRARDINIA DIVERSIFOLIA. 1, flowering stem with leaf; 2, fused pair of stipules; 3, detail of leaf from above; 4, detail of leaf from below, 1–4 *Biegel* 2889; 5, male flower, *Letouzey* 6922; 6, female flower, *Chiovenda* 1563; 7, mature infructescence, *Biegel* 2889; 8, detail of infructescence with two female flowers, *Jacques-Felix* 2690. Drawn by Victoria C. Friis. From F.T.E.A.

Distr., Thyolo (Cholo) Mt., 1200 m., female fl. & fr. 24.ix.1946, *Brass* 17778 (PRE; SRGH). **Mozambique**. T: Angónia, Mt. Dómuè, 1450 m., male fls. 9.iii., *Torre & Paiva* 11089 (LISC). MS: Inhamitanga, fl. 17.vi.1963, *M.F. Carvalho* 626 (LMU).

Widespread in the mountains of tropical Africa, from Senegal to S. Sudan and Ethiopia and south to Angola and S. Africa (Transvaal); also in Madagascar and the Yemen, and Asia from Sri Lanka and India to S. China, Taiwan and Indonesia. In montane forest, especially in clearings or along roads, in moist rocky places and moderately shaded areas near cultivation, and in caves and rocky outcrops; 1000–1600 m.

Leaf morphology of this species is extremely variable, and this has given rise to an extensive synonymy (Kew Bull. **36**: 154–155, fig. 4 (1981)). The leaves often vary considerably within a single individual.

4. **LAPORTEA** Gaudich.

Laportea Gaudich. in Freyc., Voy. Monde, Bot.: 498 (1830). —Chew in Gard. Bull.,
          Singapore **25**: 111–178 (1969) nom. conserv.
Fleurya Gaudich. in Freyc., Voy. Monde, Bot.: 497 (1830).

Annual or perennial herbs (shrubs outside the Flora Zambesiaca area), monoecious or dioecious with numerous stinging hairs (the hairs sometimes mounted on epidermal protuberances). Cystoliths linear or punctiform. Leaves alternate, petiolate, simple, serrate. Stipules intrapetiolar, fused almost to the apex. Inflorescences mostly unisexual (occasionally a few flowers of opposite sex intermixed), paniculate, sometimes apparently spicate due to reduction of branches, sometimes appearing as interrupted spikes due to elongation of some internodes. Male flowers usually pedicellate, regular, 4–5-merous, stamens equalling tepals in number, rudimentary ovary present. Female flowers pedicellate, pedicel sometimes winged; tepals 4, unequal, lateral pair usually much larger than the median pair, and median pair of unequal size; staminodes absent; ovary asymmetrical, ovoid, laterally compressed; stigma sessile, filiform, or deeply trifid with filiform branches. Achene compressed, often with a characteristic sculpturing on the sides (consisting of an annular ridge enclosing a ± rugose surface), stipitate and shed separately from the persistent perianth (Sect. *Laportea*), or sessile and shed with the persistent perianth (Sect. *Fleurya*).

A genus of about 50 species, pantropical but extending into the temperate regions of N. America and E. Asia.

This genus contains some of the most fiercely stinging plants in the family. The species are distinct but sometimes difficult to identify due to variation in leaf morphology and indumentum; however, these characters plus floral characters provide the best features for the identification of the species.

1. Stigma trifid (the central stigmatic branch considerably longer than the other two); inflorescences (at least the terminal part) spicate, apparently interrupted due to reduction of lateral branches (lowermost branches sometimes developed)   -   -   -   -   -   -   -   -   - 2
– Stigma simple; inflorescences clearly branched   -   -   -   -   -   -   -   - 3
2. Plant annual, without stolons; inflorescences usually bisexual, always axillary; stigmatic branches less than 1 mm. long; male perianth with 4 corniculate tepals   -   -   - 1. *interrupta*
– Plant perennial, usually with stolons; inflorescences unisexual, often produced directly from the stolons (female prostrate or geocarpic); stigmatic branches 2–3 mm. long; male perianth with 5 non-corniculate tepals   -   -   -   -   -   -   -   -   -   -   - 2. *ovalifolia*
3. Leaves with large teeth, margins deeply dentate with teeth usually 7–20 mm. long   -   -   -   -   -   -   -   -   -   -   -   - 3. *mooreana*
– Leaves with smaller teeth, margins dentate or serrate with teeth not exceeding 5 mm. in length   -   -   -   -   -   -   -   -   -   -   -   -   -   - 4
4. Plants densely covered with glandular hairs of varying length   -   -   4. *aestuans*
– Plants without glandular hairs   -   -   -   -   -   -   -   -   - 5
5. Pedicels of female flowers laterally winged (especially pronounced in fruit)   -   -   -   -   -   -   -   -   -   -   -   - 5. *alatipes*
– Pedicels of female flowers dorsi-ventrally winged (especially pronounced in fruit)   -   -   -   -   -   -   -   -   -   -   - 6. *peduncularis*

1. **Laportea interrupta** (L.) Chew in Gard. Bull., Singapore **21**: 200 (1965). —Letouzey in Fl.
          Cameroun 8: 125 (1968). —Chew in Gard. Bull., Singapore **25**: 145 (1969). TAB. **26**, fig. 1. Type
          from India.
          Urtica interrupta L., Sp. Pl.: 985 (1753). Type as above.

Boehmeria interrupta (L.) Willd., Sp. Pl. **4**, 1: 342 (1805). Type as above.

Fleurya interrupta (L.) Gaudich. in Freyc., Voy. Monde, Bot.: 497 (1830). —Rendle in F.T.A. **6**, 2: 248 (1917). —Hauman in F.C.B. **1**: 190 (1948). —Agnew, Upland Kenya Wild Fl.: 321 (1974). Type as above.

Urtica lomatocarpa Steud. in Flora **33**: 260 (1850). —A. Rich., Tent. Fl. Abyss. **2**: 261 (1851). Type from Ethiopia.

Annual herbs up to c. 1 m. tall, monoecious; stems sparsely branching. Stinging hairs few, up to c. 1.5 mm. long, mostly restricted to upper part of the stem. Leaves crowded towards stem apex, 8–12 × 5–7 cm., ovate; apex acuminate; base rounded to truncate; margin finely serrate with 12–15(30) teeth per side; lamina glabrous or puberulent on both sides, with few stinging hairs, lateral nerves 3–6 pairs. Petiole 5–8 cm. long, pilose to glabrescent, stinging hairs occasional. Stipules 3–5 mm. long, linear. Inflorescences usually bisexual, in axils of upper leaves, or in axils of fallen leaves just above soil level, paniculate, but apparently interrupted-spicate due to reduction of side branches, up to c. 30 cm. long, flowers in cymose clusters c. 10 mm. in diam. Male flowers appearing first, pedicels c. 1 mm. long, perianth 1–2 mm. in diam., (3)4-merous, tepals corniculate. Female flowers pedicellate, pedicels c. 0.5 mm. long, unwinged; ovary compressed, stigma sessile 3-branched. Achenes c. 1.75 mm. long, ovoid, laterally compressed, with a circular ridge surrounding a rugose central part, not stipitate, shed without the perianth.

Mozambique. MS: Inhamitanga, female fl. 7.v.1942, *Torre* 4085 (LISC).

Almost pantropical. Evergreen rainforest and riverine forest at low altitudes; weed of cultivation in partial shade, or along roadsides in forest.

The cited specimen represents the only record from the Flora Zambesiaca area so far.

2. **Laportea ovalifolia** (Schumach.) Chew in Gard. Bull., Singapore **21**: 201 (1965). —Letouzey in Fl. Cameroun **8**: 131 (1968). —Chew in Gard. Bull., Singapore **25**: 149 (1969). —Friis in F.T.E.A., Urticaceae: 18 (1989). TABS. **25**, fig. 5; **26**, fig. 2. Type from Ghana.

Haynea ovalifolia Schumach. in Schumach. & Thonn., Beskr. Guin. Pl.: 406 (1827). Type as above.

Fleurya podocarpa Wedd. in DC., Prodr. **16**, 1: 76 (1869). —Rendle in F.T.A. **6**, 2: 251 (1917). —Hauman in F.C.B. **1**; 191, t. 20 (1948) —Robyns, Fl. Parc Nat. Alb. **1**: 74 (1948). Type from Nigeria.

Fleurya podocarpa var. *mannii* Wedd. in DC., Prodr. **16**, 1: 76 (1869). —Rendle in F.T.A. **6**, 2: 251 (1917). —Hauman in F.C.B. **1**: 192 (1948). Type from Cameroon.

Fleurya podocarpa var. *amphicarpa* Engl., Pflanzenw. Ost-Afr. **C**: 163 (1895). Type from SW. Uganda or NW. Tanzania.

Fleurya podocarpa var. *fulminans* Hiern, Cat. Afr. Pl. Welw. **1**, 4: 989 (1900). Type from Angola.

Fleurya podocarpa subsp. *repens* Hauman in F.C.B. **1**: 192 (1948) nomen. Type from Zaire.

Fleurya ovalifolia (Schumach.) Dandy in F.W. Andr., Fl. Pl. Anglo-Egypt. Sudan **2**: 277 (1952). —Keay in F.W.T.A. ed. 2, **1**: 619 (1958). —Agnew, Upland Kenya Wild Fl.: 321 (1974). Type as above.

Fleurya ovalifolia var. *repens* (Hauman) Lambinon in Bull. Soc. Roy. Bot. Belg. **91**: 200 (1959). Type as above.

Perennial stoloniferous herbs, monoecious, with scattered stinging hairs, the main stem often prostrate with erect shoots, up to 2 m. tall, slightly branched. Leaves 8–10 × 4–6 cm., ovate; apex acute to acuminate; base usually subcordate to rounded; margin crenate to serrate; stinging hairs 0–few on upper surface, more numerous on lower surface, especially on the midribs. Petioles 5–10 cm. long. Stipules up to 1 cm. long, almost glabrous except for a few stinging hairs. Inflorescences unisexual. Male inflorescences often arising directly from the stolons, but sometimes also axillary from the erect stems, paniculate, up to 50 cm. long; male flowers gathered in dense clusters at intervals along the axis, pubescent, rarely with stinging hairs. Female inflorescence mostly axillary, but sometimes produced directly from the stolons, shortly racemose or paniculate, up to 6 cm. long, pubescent. Male flowers on pedicels 1–2 mm. long; perianth up to 2 mm. long, tepals 5, non-corniculate, stamens 5, pistillode 1. Female flowers on pedicels 1 mm. long; perianth up to 2 mm. long, tepals 4 almost equal, glabrous; stigma with 3 branches, a central longer one and 2 shorter lateral ones. Achene up to 3 mm. long, rugose; geocarpic fruits may occur.

Zimbabwe. N: Makonde Distr., Mhangura (Mangula), 1300 m., male fl. 25.iii.1967, *Jacobsen* 3134 (PRE; SRGH).

Widespread in tropical Africa; also from Sierra Leone to S. Sudan and south to Angola and Tanzania. Riverine forest.

The specimen cited represents the only record from the Flora Zambesiaca area so far.

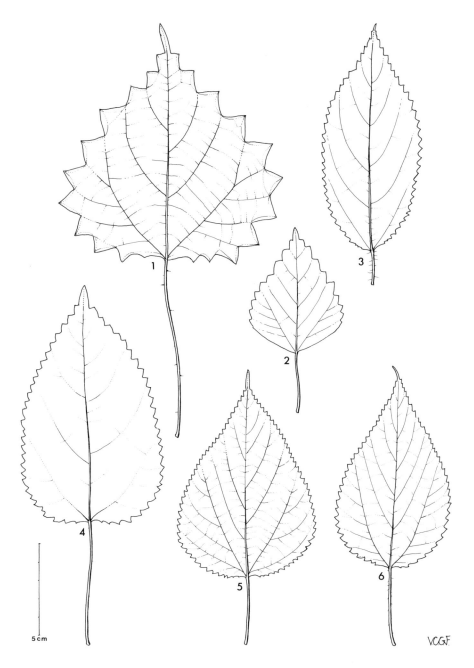

Tab. 25. LAPORTEA spp. Leaf shapes. 1.—L. MOOREANA, *Salubeni* 554; 2.—L. PEDUNCULARIS subsp. LATIDENS, *Mogg* 32147; 3.—L. ALATIPES, *Brass* 10572; 4.—L. AESTUANS, *Pawek* 9533; 5.—L. OVALIFOLIA, *Jacobsen* 3134; 6.—L. PEDUNCULARIS subsp. PEDUNCULARIS, *Müller* 3398. Drawn by Victoria C. Friis. From F.T.E.A.

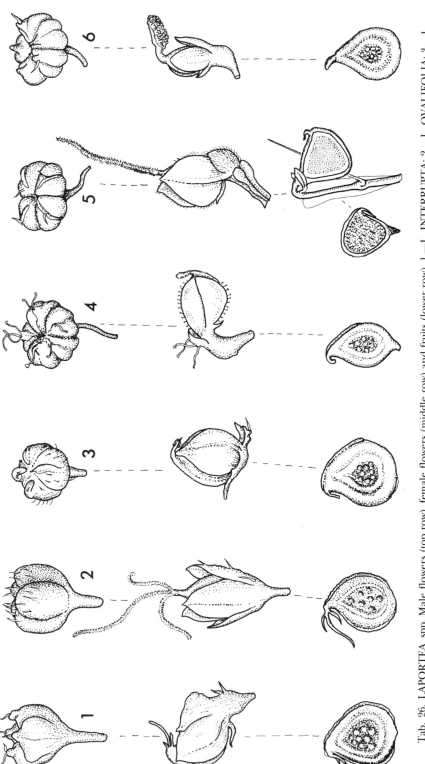

Tab. 26. LAPORTEA spp. Male flowers (top row), female flowers (middle row), and fruits (lower row). 1.—L. INTERRUPTA; 2.—L. OVALIFOLIA; 3.—L. MOOREANA; 4.—L. AESTUANS; 5.—L. ALATIPES; 6.—L. PEDUNCULARIS subsp. PEDUNCULARIS. All redrawn approximately to scale from figures published by Chew in Gard. Bull., Singapore **25**, 1: 111–178 (1969), except pedicel and fruiting perianth of 5 which is redrawn from Flore du Cameroun **8**: Pl. 18, 7. From F.T.E.A.

3. **Laportea mooreana** (Hiern) Chew in Gard. Bull., Singapore **21**: 201 (1965). —Letouzey in Fl.
Cameroun **8**: 127 (1968). —Chew in Gard. Bull., Singapore **25**: 157 (1969). —Friis in F.T.E.A.,
Urticaceae: 21 (1989). TABS. **25**, fig 1; **26**, fig 3. Type from Angola.

   Adicea mooreana Hiern, Cat. Afr. Pl. Welw. **1**, 4: 991 (1900). Type as above.

   Pilea mooreana (Hiern) K. Schum. in Just, Bot. Jahresb. **28**: 463 (1900). Type as above.

   Fleurya mooreana (Hiern) Rendle in F.T.A. **6**, 2: 250 (1917). —Keay in F.W.T.A. ed. 2, **1**: 619
   (1958). Type as above.

   Fleurya funigera Mildbr. in Notizbl. Bot. Gart. Berl. **8**: 278 (1923). Type from Cameroon.

   Fleurya urophylla Mildbr. in Notizbl. Bot. Gart. Berl. **8**: 279 (1923). —Hauman in F.C.B. **1**: 189
   (1948). Type from E. Zaire.

   Laportea grossa sensu auct. non (Wedd.) Chew (1965). —Chew in Gard. Bull., Singapore **25**:
   159 (1969) pro parte quoad specm. Mocamb.

Annual herbs, monoecious. Stems erect, few from a ± woody base, to c. 1 m. tall, sparsely
branched, covered with long stinging hairs borne on long thin protuberances, otherwise
glabrous. Leaves 4.5–15 × 3.5–13 cm., broadly ovate to triangular; apex acuminate to
caudate; base truncate to subcordate; margin grossly toothed, teeth c. 10–18 per side, each
tooth up to 1.5 × 0.7(1.5) cm.; lamina triplinerved, upper surface with scattered,
unmounted stinging hairs, lower surface with mounted stinging hairs on the nerves.
Petiole (1.5)5–15 cm. long, distal part usually densely beset with mounted stinging hairs.
Stipules up to 1 cm. long, lanceolate, fused for ± half their length, glabrescent or with a few
stinging hairs. Inflorescences unisexual or bisexual, paniculate, up to 30 × 10 cm., axillary
on peduncles 2–5 cm. long. Male flowers in separate inflorescences in the lower leaf axils,
or in the lower part of bisexual inflorescences on pedicels c. 1 mm. long; perianth c. 1 mm.
long, tepals 4 ± corniculate, stinging hairs few; stamens 4, pistillode small. Female flowers
in inflorescences of upper leaf axils on pedicels 1–2 mm. long; tepals 3 or 4, unequal, with
a few stinging hairs. Achene c. 1.5 mm. long, ovoid, laterally compressed, sessile to slightly
stipitate, rugose on the flattened sides, shed with the perianth.

   Zimbabwe. E: Vumba Mt., Leopard's Rock Hotel, 1600 m., male fl. 4.i.1974, *Bamps, Symoens & van
den Berghen* 559 (BR; SRGH). **Malawi**. C: Dedza Distr., Domwe Hill, 2300 m., male fl. 1.iv.1961,
Chapman 1218 (SRGH). S: Zomba Plateau, 1050 m., male fl. 6.iii.1977, *Brummitt & Seyani* 14811 (K;
SRGH). **Mozambique**. N: Cabo Delgado, 500 m., female fl. 3.iv.1964, *Torre & Paiva* 11723
(LISC). Z: Milange, Serra Chiperone, 1150 m., fl. 30.i.1972, *Correia & Marques* 2401 (LMU). T:
Zóbuè, male & female fls. 3.x.1942, *Mendonça* 598 (LISC).

   Widely distributed in tropical Africa; also from Nigeria to Uganda and south to Angola.
Understorey of mixed evergreen rainforest, often in rocky places; 900–1600 m.

   A fiercely stinging plant.

4. **Laportea aestuans** (L.) Chew in Gard. Bull., Singapore **21**: 200 (1965). —Letouzey in Fl. Cameroun
8: 121 (1968). —Chew in Gard. Bull., Singapore **25**: 164 (1969). —Friis in F.T.E.A., Urticaceae: 23
(1989). TABS. **25**, fig. 4; **26**, fig. 4. Type from (?)Surinam.

   Urtica aestuans L., Sp. Pl., ed. 2: 1397 (1753). Type as above.

   Urtica hirsuta Vahl, Symb. Bot. **1**: 77 (1790). Type from the Yemen.

   Fleurya aestuans (L.) Gaudich. in Freyc., Voy. Monde, Bot.: 497 (1830). —Rendle in F.T.A. **6**, 2:
   246 (1917). —Hauman in F.C.B. **1**: 188 (1948). —Keay in F.W.T.A., ed. 2, **1**: 619 (1958). —Agnew,
   Upland Kenya Wild Fl.: 321 (1974). Type as above.

   Urtica schimperiana Steud. in Flora **33**: 259 (1850). —A. Rich., Tent. Fl. Abyss. **2**: 261 (1851).
   Type from Ethiopia.

   Fleurya aestuans var. *linnaeana* Wedd. in DC., Prodr. **16**, 1: 72 (1969). Type as for typical
   variety.

   Fleurya perrieri Léandri in Ann. Mus. Col. Marseille, Sér. 6, 7–8: 16, fig. 3 (1950) non *Laportea
   perrieri* J. Léandri (1950). Type from Madagascar.

   Laportea bathiei Léandri in Fl. Madag. **56**, Urtic.: 10, t. 2, fig. 1–7 (1965). Type as for *Fleurya
   perrieri* J. Léandri.

Erect annual monoecious herbs to c. 1 m. tall. Stems somewhat fleshy above, ± woody
below, little branched, leafy in upper part, covered with stinging hairs to c. 1 mm. long and
soft glandular hairs 1–2(3) mm. long. Leaves 10–15 × 8–12 cm., ovate to broadly ovate;
apex acuminate; base rounded to truncate; margin regularly serrate, with 30–40 teeth per
side c. 0.5 mm. long; lamina usually with stinging hairs on both sides but without
glandular hairs. Petiole 5–15 cm. long, usually covered with soft glandular hairs and a few
stinging hairs. Stipules up to 1 cm. long, linear-lanceolate, fused for ± half their length.
Inflorescences bisexual, mainly in the axils of the uppermost leaves, consisting of
profusely branched panicles up to c. 20 cm. long, on peduncles 4–10 cm. long, axes
covered with glandular and stinging hairs. Male flowers intermixed with female flowers
or mostly at the base of the inflorescences, soon falling, pedicellate,

perianth 4–5-merous, 1–2 mm. in diam., usually with glandular hairs on the outside. Female flowers densely clustered, tepals 4 unequal, pedicel unwinged. Achenes 1–2 mm. long, ovoid, laterally compressed, with circular and rugose markings on the sides, shortly stipitate, shed without the perianth.

Zambia. C: Luangwa Valley, 700 m., female fl. 18.iv.1966, *Astle* 4783 (SRGH). S: Sinazongwe, male fl. 28.v.1961, *Fanshawe* 6625 (SRGH). **Zimbabwe.** N: Hurungwe Distr., Zambezi Valley, Kariba Gorge, fr. 25.ii.1953, *Wild* 4092 (PRE; SRGH; UPS). **Malawi.** N: Karonga, 500 m., fl. 24.iv.1975, *Pawek* (K; MO; PRE; SRGH). S: Ntchisi (Chilwa) Isl., 760 m., female fl. 11.iii.1955, *Exell, Mendonça & Wild* 811 (SRGH). **Mozambique.** N: Nampula Prov., Malema (Entre Rios), 550 m., male & female fls. 13.ii.1964, *Torre & Paiva* 10562 (LISC). T: Tete, Cabora Bassa, fl. & fr. 21.ii.1972, *Macedo* 4879 (LISC; LMU). MS: Chimoio, Serra de Chindaza, male & female fls. 26.iii.1948, *Garcia* 738 (LISC).

Pantropical. Weed in partial shade along roads and in disturbed areas in forest and scrub, and in cultivation; 250–1000 m.

An unpleasantly stinging plant.

5. **Laportea alatipes** Hook. f. in Journ. Linn. Soc., Bot. **7**: 215 (1864). —Wedd. in DC., Prodr. **16**, 1: 79 (1869). —Rendle in F.T.A. **6**, 2: 252 (1917). —Hauman in F.C.B. **1**: 194 (1948). —Robyns, Fl. Parc Nat. Alb. **1**: 71, t. 6 (1948). —Keay in F.W.T.A., ed. 2, **1**: 620 (1958). —Letouzey in Fl. Cameroun **8**: 117, t. 18 (1968). —Chew in Gard. Bull., Singapore **25**: 124 (1969). —Agnew, Upland Kenya Wild Fl.: 321 (1974). —Friis in F.T.E.A., Urticaceae: 16 (1989). TABS. **25**, fig. 3; **26**, fig. 5. Type from Cameroon.

Urticastrum alatipes (Hook. f.) Kuntze, Rev. Gen. Pl. **2**: 635 (1891). Type as above.
Fleurya urticoides Engl., Bot. Jahrb. **33**: 122 (1902). Types from Cameroon.
Girardinia marginata Engl., Bot. Jahrb. **33**: 123 (1902). Type from Cameroon.
Fleurya urticoides var. *glabrata* Rendle in F.T.A. **6**, 2: 248 (1917). Type from Tanzania.
Fleurya alatipes (Hook. f.) N.E. Br. in F.C. **5**, 2: 547 (1925).

Annual or short-lived perennial herbs to 1.5 m. tall, monoecious, with numerous stinging hairs on all aerial parts, but without glandular hairs; stems soft but somewhat woody at the base, erect or decumbent. Leaves usually fallen from lower part of the stem at anthesis, 8–15 × 4–8 cm., ovate; apex acuminate; base cuneate, rounded or subcordate; margin coarsely serrate with 25–35 teeth per side less than 5 mm. long; upper surface of lamina with stinging hairs and a few stiff hairs, lower surface pubescent. Petiole 5–10(12) cm. long. Stipules up to 1.5 cm. long, lanceolate, fused for more than half of their length. Inflorescences unisexual, much branched panicles; male inflorescences up to 10 cm. long, usually in the axils of the lower leaves; female inflorescences up to 20 cm. long, usually in the axils of the upper leaves, on peduncles 4–9 cm. long. Male flowers pedicellate, in clusters 2–3 mm. in diam., perianth 4-merous, mostly with stinging hairs. Female flowers with broadly winged pedicels, wings broadening to c. 1.3 mm. in fruit; flowers in fan-shaped clusters, perianth tepals unequal, stigma filiform. Achene slightly stipitate, c. 1 mm. long, laterally compressed with the centre of the flattened sides irregularly rugose, shed with the perianth.

Zambia. E: Nyika Plateau, Chowo Forest, 2000 m., female fls. 14.iv.1991, *Friis* 6322 (C; K; MAL). **Zimbabwe.** E: Engwa, Himalayas, 2200 m., fl. 2.iii.1954, *Wild* 4437 (PRE; SRGH; UPS). **Malawi.** N: Rumphi Distr., Nyika Nat. Park, Kaziyula Forest, fr. 6.iv.1981, *Salubeni & Tawakali* 3001 (SRGH). C: Ntchisi Mt., 1450 m., male & female fls. 19.ii.1959, *Robson* 1675 (K; PRE; SRGH). S: Mulanje Mt., Luchenya Plateau, fr. 1.vii.1946, *Brass* 16572 (K; SRGH).

Widespread, in the mountains of Cameroon, E. Zaire, Ethiopia, Uganda, Kenya, Tanzania and south to S. Africa (Transvaal). In understorey of montane forests and forests in ravines; 1400–2200 m.

A fiercely stinging plant. It may be distinguished from *L. aestuans* by that species having long soft glandular hairs, from *L. peduncularis* by that species having 5-merous male flowers and from *L. interrupta* by that species having long bisexual inflorescences and branched styles.

6. **Laportea peduncularis** (Wedd.) Chew in Gard. Bull., Singapore **21**: 201 (1965); in Gard. Bull., Singapore **25**: 152 (1969). —Friis in Bol. Soc. Brot., sér. 2, **58**: 205 (1985); in F.T.E.A., Urticaceae: 19 (1989). Type from S. Africa (Natal).

Annual or short-lived perennial, erect or ascending herbs, monoecious or dioecious by abortion. Stems soft, ± erect, decumbent or scrambling, sometimes rooting at lower nodes, up to 1.5 m. long, unbranched or branching from the bases, glabrescent or covered with few to numerous stinging hairs, sometimes on raised protuberances. Leaves restricted to upper part of stem at anthesis, up to 10 × 7 cm., triangular to rhomboid or ovate; apex acuminate; base cuneate, truncate or subcordate; margin coarsely to finely serrate with 5–25 teeth per side less than 5 mm. long; lamina upper surface with short, stiff hairs or subglabrous, lower surface with stiff hairs on the nerves or subglabrous. Petiole 2–6(11) cm. long, glabrous to puberulous. Stipules up to 10 mm. long, linear-lanceolate, fused for about half their length, with a few stiff hairs. Inflorescences single and usually restricted to upper leaf axils, unisexual, paniculate, usually dichotomously branched, with many flowers in small cymose clusters or with a few flowers on each branch, with scattered stiff or stinging hairs. Male inflorescences in lower axils, on peduncles 1–3 cm. long; flowers in few, dense cymose clusters c. 1 cm. in diam.; pedicels c. 1 mm. long with a narrow dorsi-ventral wing; perianth with 5 strongly corniculate tepals, 5 stamens, and a small pistillode. Female flowers in small dichasial clusters c. 6 mm. in diam.; pedicels c. 1 mm. long with a narrow dorsi-ventral wing; tepals 4, c. 1 mm. long, the dorsal and ventral ones much smaller than the lateral ones. Achene c. 1.5 mm. long, ovoid, laterally compressed, with a ridge enclosing a rugose depression on each side, shed with the perianth.

The nomenclature of this species is somewhat confusing, as the earliest name used for it in *Fleurya* (now considered congeneric with *Laportea*) is *Fleurya capensis* (used by Weddell in 1854 and 1856) which is presumably based, through an indirect reference by Thunberg, on a name (*Urtica capensis*) proposed by the younger Linnaeus. This name is, according to the original material referable to the genus Didymodoxa.
The complex nomenclatural situation has been discussed in more detail by Friis in Bol. Soc. Brot. **58**: 208–9 (1985) and Friis and Wilmot-Dear in Nordic Journ. Bot. **8**: 47 (1988).
L. peduncularis is a polymorphic species, within which two distinct subspecies can be recognised, both occurring in the Flora Zambesiaca area. The two subspecies are not only morphologically but also ecogeographically quite distinct, with subsp. *peduncularis* in montane forest and scrub from N. Tanzania to S. Africa (Cape Province), and subsp. *latidens* in coastal forest and scrub from Mozambique and S. Africa (Natal).

Lamina ovate, with a truncate or cordate base and with more than 12 fine, narrow teeth on each half
  of the lamina; plant creeping, scrambling or suberect, rarely small and erect, subglabrous or
  hispid, often with mounted stinging hairs   -   -   -   -   -   -   - subsp. *peduncularis*
Lamina rhomboid to ovate, with a cuneate base and with 5–8(12) broad, blunt teeth on each half of
  the lamina; plant more or less erect, usually glabrous, and without stinging
  hairs   -   -   -   -   -   -   -   -   -   -   -   -   -   - subsp. *latidens*

Subsp. **peduncularis** TABS. **25**, fig. 6; **26**, fig. 6.
    Fleurya mitis Wedd. in Ann. Sci. Nat., Bot. Sér. 4, **1**: 183 (1854) nom. nud.
    Fleurya capensis sensu Wedd. in Archiv. Mus. Nat. Hist. Nat., Paris 9 (Monogr. Urtic.): 117, t. 1,
    fig. A, 7–8 (1856). —Rendle in F.T.A. **6**, 2: 249 (1917), non. *F. capensis* (L.f.) Wedd. (1854), as "F. c.
    (Thunb.) Wedd."
    Fleurya capensis var. *mitis* Wedd., in Archiv. Mus. Nat. Hist. Nat., Paris 9 (Monogr. Urtic.): 118
    (1856). Syntypes from S. Africa (Natal).
    Fleurya peduncularis Wedd. in DC., Prodr. **16**, 1: 75 (1869). Type as above.
    Fleurya peduncularis var. *mitis* (Wedd.) Wedd. in DC., Prodr. **16**, 1: 76 (1869). Types as above.
    Fleurya mitis (Wedd.) N.E. Br. in F.C. **5**, 2: 546 (1925). Types as above.
    Laportea caffra Chew in Gard. Bull., Singapore **25**: 155 (1969). Type from S. Africa (Cape
    Province).

Annual herbs. Stems often prostrate or scrambling, sparsely branched, erect branches up to c. 1 m. long, younger parts covered with stinging hairs on short protuberances, rarely glabrescent. Leaves 6–10 × 3–7 cm., ovate; apex acuminate; base truncate, rounded or subcordate; margin finely serrate or toothed, teeth up to 25 on each half of the lamina and usually less than 5 mm. high; lamina upper surface with short stiff hairs, lower surface with short stiff hairs on the nerves. Petioles 4–6 cm. long. Stipules up to 7 mm. long, with a few stiff hairs. Inflorescences sparsely branched, axillary, with scattered stiff or stinging hairs; male 4–6 × 3–5 cm.; female 4–6 × 2–3 cm.

Zimbabwe. E: Mutare, Range Hill, Commonage, 1200 m., fl. 12.i.1960, *Chase* 7259 (PRE; SRGH). **Malawi**. N: Chitipa Distr., Kasumbi Thicket, Misuku Hills, female fl. 15.iv.1981, *Salubeni & Tawakali* 3117 (SRGH). C: Ntchisi Mt., male fl. 14.iii.1967, *Salubeni* 596 (SRGH). S: Thyolo (Cholo) Mt., 1200 m., female fl. 22.ix.1946, *Brass* 17733 (BR; K; MO; PRE; SRGH). **Mozambique**. MS: Chimoio, Serra de Garuso, male & female fls. 27.iii.1948, *Barbosa* 1243 (LISC). GI: Gaza, fl. ii.1948, *Torre* 7287 (LISC).
    Also in Tanzania, S. Africa (Transvaal, Natal and Cape Province), and Swaziland. In montane and upland evergreen forest and montane scrub, along paths and rivers; 1100–1700 m. In S. Cape Province (Knysna) in forest near sea level.

Subsp. **latidens** Friis in Bol. Soc. Brot. **58**: 206 (1985). TABS. **25**, fig. 2; **27**. Type: Mozambique, Inhaca Island, near sea level, 8.ix.1964, *Mogg* 30920 (K, holotype; LMU, isotype).

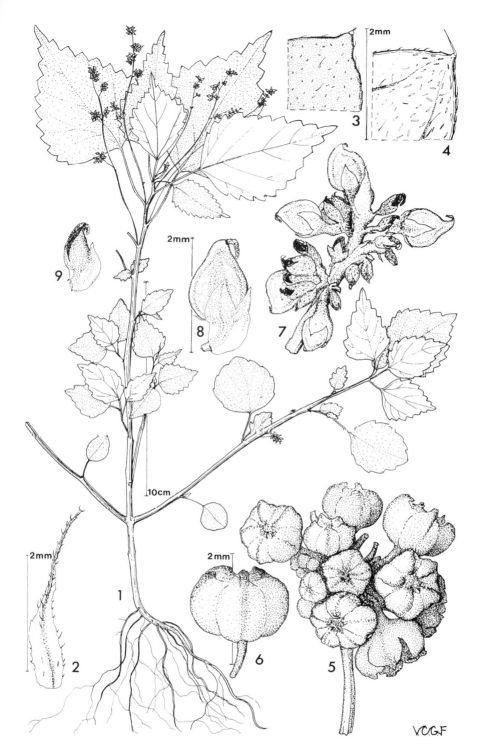

Tab. 27. LAPORTEA PEDUNCULARIS subsp. LATIDENS. 1, habit; 2, stipule; 3, leaf, superior surface, detail; 4, leaf, inferior surface, detail, 1–4 *Mogg* 32147; 5, part of male inflorescence; 6, male flower, bud, 5–6 *Barbosa & de Lemos* 8401; 7, part of female inflorescence; 8, older female flower; 9, young female flower, 7–9 *Mogg* 32147. Drawn by Victoria C. Friis.

Erect, more rarely ascending, annual herbs, up to 45 cm. tall. Stems and petioles glabrous to subglabrous. Leaf 4.0–7.0 × 2.5–4.6 cm., rhomboid to ovate; apex acuminate; base cuneate to subtruncate; margin coarsely serrate, with 5–8(12) broadly triangular teeth on each half of lamina; lamina upper surface with very few, scattered stiff hairs, lower surface subglabrous. Male inflorescence usually regularly dichotomously branched, or sometimes with one division overtopping the other, flowers in clusters of 10–20.

Mozambique. GI: Gaza, Bilene, Chissano, male fl., 13.ii.1959, *Barbosa & Lemos* 8401 (COI; K; LISC; PRE; SRGH). M: Inhaca Isl., 400 m., male & female fls. 30.ix.1959, *Mogg* 32147 (LMU; SRGH).
Also in the coastal zone of S. Africa (Natal). In dune forest and coastal scrub, apparently never very far from the sea; 50–400 m.

5. PILEA Lindl.

Pilea Lindl., Coll. Bot. t. 4 (1821). —Friis in Kew Bull. **44**: 557–600 (1989) nom. conserv.

Annual or perennial herbs, monoecious or dioecious by abortion. Stems juicy, translucent and turgescent; stinging hairs absent. Leaves opposite, petiolate or subsessile, each of a pair unequal in size, triplinerved, usually with serrate margins. Stipules intrapetiolar, almost completely fused. Cystoliths linear. Flowers in unisexual or bisexual cymes or panicles. Male flowers (3)4(5)-merous, tepals corniculate with a dorsal horn-like appendage; stamens as many as the tepals; ovary rudimentary. Female flowers with 3 tepals, one considerably larger than the other two, usually cuculate or with a dorsal horn-like appendage. Three scale-like staminodes present, inflexed, but ultimately reflexing and thus ejecting the achene. Ovary symmetrical, erect, with a sessile, penicillate stigma. Achene ovoid, ± smooth, compressed. Embryo with large cotyledons.

A genus of about 200 species, almost pantropical in distribution and in most subtropical parts of the world with the exception of Australia and New Zealand.
Pilea microphylla (L.) Liebm., a species with very small, 1–8 mm. long, entire leaves and a native of tropical America, is cultivated as an ornamental garden plant or occurs as a garden weed in E. Africa and the Flora Zambesiaca area, *Friis* 6328, Malawi, Mzuzu, garden (C; K; MAL).
Some of the species are very polymorphic, and others difficult to distinguish.

1. Female inflorescences restricted to the top of the plant, consisting of very dense cymes subtended by the four uppermost leaves in an apparent whorl - - - - - 1. *tetraphylla*
   - Female inflorescences distributed along the stems in axils of leaf pairs separated by clearly discernible internodes, not restricted to the axils of the four uppermost leaves - - 2
2. Inflorescences of sessile heads, clustered, forming apparent whorls at the nodes - - - - - - - - - - - - - - 2. *rivularis*
   - Inflorescences of clearly pedunculate heads or of ± interrupted spike-like panicles, not forming apparent whorls at the nodes - - - - - - - - - - - 3
3. Stipules persisting; leaves 2–6 × 1.5–4.5 cm., with 7–15 coarse teeth on each margin; plants erect, unbranched or little branched herbs with thick, fleshy stems up to 40 cm. tall - - - - - - - - - - - - - - - 3. *johnstonii*
   - Stipules early deciduous; leaves 1–3 × 0.8–2.5 cm., with 4–9 teeth on each margin; plants prostrate or erect, ± branched herbs with thin stems up to 10(30) cm. tall - - - - - - - - - - - - - 4. *usambarensis*

1. **Pilea tetraphylla** (Steud.) Blume, Mus. Bot. Lugd.-Bat. **2**: 50 (1856). —Wedd. in DC., Prodr. **16**, 1: 136 (1869). —Rendle in F.T.A. **6**, 2: 270 (1917). —Hauman in F.C.B. **1**: 199 (1948). —F. W. Andr., Fl. Pl. Anglo-Egypt. Sudan **2**: 279 (1952). —Keay in F.W.T.A., ed. 2, **1**: 621 (1958). —Léandri in Fl. Madag. **56**, Urtic.: 44, t. 9, fig. 5–7 (1965). —Letouzey in Fl. Cameroun **8**: 173 (1968). —Agnew, Upland Kenya Wild Fl.: 323 (1974). —Friis in F.T.E.A., Urticaceae: 27 (1989). TAB. **28**, figs. A1–A5. Type from Ethiopia.
   Urtica tetraphylla Steud. in Flora **33**: 260 (1850). Type as above.
   Pilea quadrifolia A. Rich., Tent. Fl. Abyss. **2**: 263 (1850). —Wedd. in Archiv. Mus. Nat. Hist. Nat., Paris 9 (Monogr. Urtic.): 199 (1856). Type from Ethiopia.
   Pilea hypnopilea Bak. in Journ. Bot. **20**: 267 (1882). Type from Madagascar.
   Pilea modesta Bak. in Journ. Linn. Soc., Bot. **20**: 265 (1884). Type from Madagascar.
   Adicea tetraphylla (Steud.) Kuntze, Rev. Gen. Pl. **2**: 623 (1891). Type as above.
   Pilea tetraphylla var. *major* Rendle in Journ. Bot. **55**: 201 (1916); in F.T.A. **6**, 2: 271 (1917). Type from Cameroon.
   Pilea tetraphylla var. *hypnopilea* (Bak.) Léandri in Fl. Madag. **56**, Urtic.: 46 (1965). Type as above.

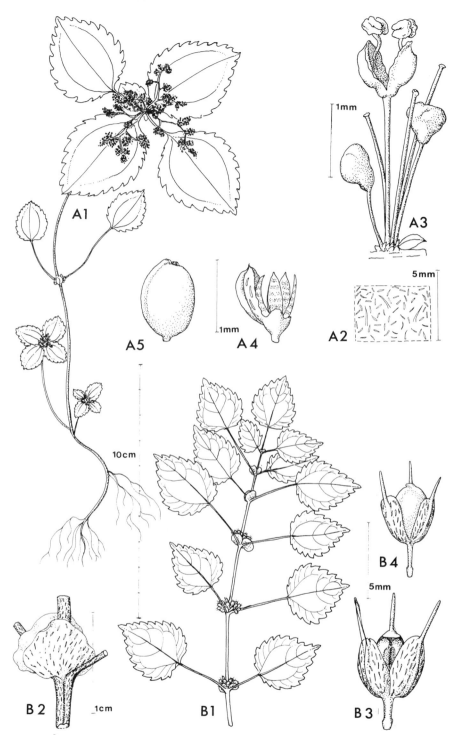

Tab. 28. A.—PILEA TETRAPHYLLA. A1, habit; A2, leaf, superior surface, detail, A1–2 *Jacques-Felix* 2740; A3, male inflorescence; A4, old female flower, showing unilateral perianth, achene ejected; A5, achene, A3–5 *Leeuwenberg* 6972. B.—PILEA RIVULARIS. B1, flowering stem, upper part; B2, stipules; B3, male flower, stamens still inflexed, B1–3 *Mann* 2011; B4, female flower, *Jacques-Felix* 2758. Drawn by Victoria C. Friis. From F.T.E.A.

Erect annual herbs, monoecious (or dioecious by abortion). Stems to c. 20 cm. tall, fleshy, pinkish, glabrous; leaves of a pair equal, increasing in size upwards, the four uppermost forming an apparent whorl at the top of the stem by shortening of the uppermost internode. Leaves petiolate, 0.7–2 × 0.4–1.5 cm., ovate, apex acute or short acuminate; base ± broadly cuneate; margin serrate with 5–12 teeth per side, lamina glabrous on both sides. Petiole 2–25 mm. long. Stipules up to 4 mm. long, broadly ovate, membranous, persistent. Male inflorescences small and insignificant, in the axils of the leaves below the apical whorl, sometimes absent, sometimes with a few female flowers, or a few male flowers in the apical inflorescence; male flowers almost sessile, 2–4-merous, with stamens in the same number as tepals. Apical inflorescence entirely or almost entirely female; female flowers almost sessile, with one large and 2(3?) smaller tepals, and the same number of staminodes, stigma sessile, penicillate. Achene c. 1.5 mm. long, broadly ovoid, compressed, smooth or ± verrucose.

Zambia. N: Mbala Distr., Tasker's Deviation, waterfall, 1650 m., male & female fls., *Richards* 1684 (K). **Zimbabwe**. E: Mutare Distr., Vumba Mt., Leopard's Rock, male & female fls., *Whellan* 1593 (SRGH). **Malawi**. N: Rumphi Distr., Nyika Nat. Park, 2300 m., male & female fls., fr. 27.iv.1973, *Pawek* 6640 (K; MO; SRGH). C: Dedza Mt., male & female fls. 1.v.1968, *Salubeni* 1081 (SRGH).
Widespread in mountain areas of tropical Africa. Also from Nigeria and Cameroon, east to Ethiopia and south to Angola; and in Madagascar. In the humid, shady understorey of montane evergreen forest, in moist places beside boulders and in rock crevices, or as a weed in gardens; 1650–2800 m.

2. **Pilea rivularis** Wedd. in Archiv. Mus. Nat. Hist. Nat., Paris 9 (Monogr. Urtic.): 266 (1856); in DC., Prodr. **16**, 1: 136 (1869). —Léandri in Fl. Madag. **56**, Urtic.: 45 (1965). —Letouzey in Fl. Cameroun **8**: 163, t. 27 (1968). —Friis in F.T.E.A., Urticaceae: 29 (1989). TAB. **28**, figs. B1–B4. Type from the Comoro Islands.
   Pilea ceratomera Wedd. in DC., Prodr. **16**, 1: 132 (1869). —Engl. in Mildbr., Wiss. Ergebn. Deutsch. Zentr.-Afr. Exped. 1907–1908, **2**: 191 (1911). —Rendle in F.T.A. **6**, 2: 269 (1917). —Hauman in F.C.B. **1**: 203 (1948). —Robyns, Fl. Parc Nat. Alb. **1**: 80 (1948). —F. W. Andr., Fl. Pl. Anglo-Egypt. Sudan **2**: 279 (1952). —Keay in F.W.T.A., ed. 2, **1**: 621 (1958). —Lambinon in Bull. Soc. Roy. Bot. Belg. **91**: 204 (1959). —Agnew, Upland Kenya Wild Fl.: 323 (1974). Type from Bioko.
   Pilea macrodonta Bak. in Journ. Linn. Soc., Bot. **20**: 266 (1884). Type from Madagascar.
   Pilea ceratomera var. *mildbraedii* Engl. in Mildbr., Wiss. Ergebn. Deutch. Zentr.-Afr. Exped. 1907–1908, **2**: 191 (1911). —Rendle in F.T.A. **6**, 2: 270 (1917). —Hauman in F.C.B. **1**: 204 (1948). Type from Rwanda.
   Pilea comorensis Engl., Bot. Jahrb. **33**: 124 (1902). Type from the Comoro Islands.
   Pilea worsdellii N.E. Br. in F.C. **5**, 2: 550 (1925). Type from South Africa (Natal).
   Pilea stipulata Hutch. & Dalz. in F.W.T.A. **1**: 443 (1928); in Bull. Misc. Inf., Kew **1929**: 19 (1929). Type from Cameroon.
   Pilea ceratomera subsp. *ceratomera* (as "ssp. *typica*"). —Hauman in F.C.B. **1**: 203 (1948). Type as for *P. ceratomera*.
   Pilea ceratomera subsp. *glechomoides* Hauman in F.C.B. **1**: 204 (1948). Type from Zaire.
   Pilea ceratomera forma *hypsophila* Hauman in F.C.B. **1**: 204 (1948). Type from Zaire.
   Pilea rivularis var. *macrodonta* (Bak.) Léandri in Fl. Madag. **56**, Urtic.: 46 (1965). Type as above.

Perennial herbs with creeping rhizomes and erect, mostly unbranched stems up to 0.5(1) m. tall, dioecious or at least each shoot unisexual. Leaves of a pair subequal or equal, 2–5(10) × 1.5–3.5(6) cm., broadly ovate to ovate-lanceolate, apex acuminate; base obtuse, truncate or rounded, rarely subcordate; margin serrate or dentate with 5–15(20) teeth per side; glabrous on both sides with dense linear cystoliths. Petiole 4–6 cm. long. Stipules up to 0.5 cm. long, broadly ovate, base often cordate or almost amplexicaul. Inflorescences of dense uni- or bisexual clusters in the leaf axils, distributed along the stems. Male and female flowers on pedicels up to 2 mm. long, 3-merous, tepals with a very prominent horn-like appendage; stigma penicillate. Achene c. 2 mm. long, ovoid, compressed, smooth.

Zambia. E: Nyika Plateau, Chowo Forest, 2000 m., male & female fls. 14.iv.1991, *Friis* 6323 (C; K; MAL). **Malawi**. N: Nkhata Bay Distr., Viphya Plateau, Mzuzu, 2100 m., male & female fls. 3.vii.1976, *Pawek* 11397 (K; MO; PRE; SRGH). S: Mt. Mulanje, Lichenya Plateau, male & female fls. 10.vi.1962, *Robinson* 5324 (M; SRGH).
Widely distributed in the moist uplands and mountains of tropical Africa. From S. Nigeria and Cameroons, eastwards to Ethiopia and south through E. Zaire and E. Africa to S. Africa (Transvaal). Also on the islands of São Tomé, Bioko, Comoro and Madagascar. High rainfall evergreen, montane forests, often in moist rocky places, ravines and along small streams; 1800–2100 m.
This is a very polymorphic species, for which a wide range of infraspecific taxa have been

proposed. I have not accepted any of these taxa here, and have followed the broad view proposed and discussed by Letouzey in Fl. Cameroun **8**: 165 (1968).

3. **Pilea johnstonii** Oliv. in Trans. Linn. Soc., ser. 2, **2**: 349 (1887). —Rendle in F.T.A. **6**, 2: 273 (1917). —Hauman in F.C.B. **1**: 200 (1948). —Agnew, Upland Kenya Wild Fl.: 323 (1974). —Friis in Kew Bull. **44**: 577 (1989); in F.T.E.A., Urticaceae: 33 (1989). Type from Tanzania.

Perennial herbs with short, creeping rhizomes and erect unbranched or little branched juicy stems up to 0.4 m. tall, monoecious. Leaves of a pair subequal, 2–6 × 1.5–4.5 cm., broadly ovate to ovate-lanceolate; apex acuminate; base cuneate, obtuse, truncate or rounded, rarely subcordate; margin coarsely serrate, with 7–15(24) teeth on each side, lamina upper surface with scattered stiff hairs and linear cystoliths, lower surface glabrous to densely pubescent, especially on the nerves. Petiole 0.3–5(10) cm. long, usually pubescent. Stipules brown, membranous, up to 5(8) mm. long, broadly ovate, with a rounded or cordate base, persisting. Inflorescences axillary, with up to 4 inflorescences at each node, flowers arranged in pedunculate heads or interrupted spikes, with 2–6 globular heads along the axis of each spike, or with a branched inflorescence axis. Male heads up to 8(12) mm. in diam. on peduncles up to 6 cm. long; male flowers on pedicels up to 2 mm. long, 4-merous, with apiculate tepals (sometimes developed into horn-like appendages). Female heads up to 6 mm. in diam., subsessile or on peduncles up to 5 cm. long; female flowers on pedicels 1–2 mm. long, 3-merous, tepals with appendages as in the male flowers, middle tepal lanceolate, up to 3 mm. long, lateral tepals shorter. Achene compressed ovoid, lanceolate, smooth, shiny, brown, up to 2 mm. long.

A rather variable species which occurs in the montane evergreen forests of S. Ethiopia, Kenya, Uganda, Rwanda, Burundi, E. Zaire, Tanzania, Malawi and Zimbabwe. Three subspecies of *Pilea johnstonii* have been recognised by Friis in Kew Bull. **44**: 577–582 (1989), the two described below plus subsp. *kiwuensis* (Engl.) Friis which is restricted to the Kiwu Province of E. Zaire.

Peduncle of the female inflorescence usually with a repeatedly branched axis, nearly always with more than two corymbose clusters of flowers, each cluster usually with a clearly visible involucre of brown bracts at the base; stipules mostly more than 3.5 mm. wide, usually long-persisting and conspicuous on most leaves; leaf lamina with scattered hairs on the upper surface - - - - - - - - - - - - - subsp. *johnstonii*
Peduncle of the female inflorescence usually with only one corymbose cluster of flowers (rarely two clusters, with one above the other in a short, apparently racemose inflorescence), each cluster with only a few, insignificant brown bracts at the base; stipules less than 3.5 mm. wide and often inconspicuous at the lower nodes (only clearly visible on the upper two pairs of leaves); leaf lamina glabrous above (or with only a few scattered hairs) - - subsp. *rwandensis*

Subsp. **johnstonii** —Friis in Kew Bull. **44**: 578 (1989); in F.T.E.A., Urticaceae: 34 (1989).
    Pilea johnstonii var. *runssorensis* Engl., Pflanzenw. Ost-Afr. **C**: 163 (1895). Type from Ruwenzori (from the border between Zaire and Uganda).

Usually a robust plant with thick, juicy stems, often reaching a height of 60 cm. or more. Leaves 1.5–8.5(12.5) × 1–7.5(8.5) cm., broadly ovate, rarely elliptic to lanceolate; apex acute to short-acuminate; base broadly cuneate, truncate or subcordate; margin serrate, with 10–24 teeth on each side; lamina with scattered hairs (or subglabrous) above, and with spreading hairs on the nerves below. Stipules membranous, usually brown and long persisting, conspicuous also at the lower nodes, 4–7.5(8.5) × 3.5–4.5 mm., ovate, apex rounded or broadly acute, base broadly cuneate to subcordate. Petiole 1.2–8(10) cm. long. Inflorescences consisting of pedunculate corymbose clusters; peduncles usually 2 in each leaf axil, frequently branched, especially the female ones. Male peduncles 1–4.5 cm. long; clusters of male flowers 8–12 mm. in diam. Female peduncles 1–3.5 cm. long, often dichotomously branched at the first cluster, and frequently with 2–3 further dichotomies; clusters of female flowers 4–8 mm. in diam.

Zimbabwe. E: Nyanga Distr., Nyamingura R. near Tea Estate, c. 1400 m., male & female fls. 24.ii.1961, *Phipps* 2858 (K; PRE; SRGH). **Malawi** S: Mt. Mulanje, Chambe-Luchenya path, 1828–2133 m., immature female fls. 20.i.1967, *Hilliard & Burtt* 4542 (E [not seen]; MAL); Mt. Mulanje, near Madzeka Hut, 1900 m., male fls. 15.iv.1991, *Thulin* 7819 (MAL; UPS).
    Also in Ethiopia, Sudan, Zaire, Uganda, Kenya and Tanzania.
    Montane evergreen forest, forest floor, often along streams; 1400–2300 m.
    In specimens from N. Kenya and Ethiopia the leaves are glabrescent or glabrous on the upper surface; in almost all specimens from further south the leaves have scattered hairs above and

numerous spreading hairs on the nerves below. Material with well developed female inflorescences has not yet been seen from the Flora Zambesiaca area; however, the material agrees well with typical material of subsp. *johnstonii* from East Africa and there is little doubt about its identity.

Subsp. **rwandensis** Friis in Kew Bull. **43**: 648 (1988); in op. cit. **44**: 581 (1989); in F.T.E.A., Urticaceae: 35 (1989). Type from E. Zaire.

Habit usually more slender than in the typical subspecies; plants rarely reaching a height of 60 cm. Leaves (1.5)3–8.5 × (1.2)1.4–4 cm., ovate to elliptic, apex acuminate, base usually cuneate rarely subcordate, margin serrate with 6–9 teeth on each side, lamina upper surface glabrous or with a few scattered hairs, lower surface with spreading hairs on the nerves. Stipules partly persistent but often difficult to see at the lower leaves, 2–3.5 × 0.8–1.2 mm., triangular. Petiole (1)1.5–4.5(6) cm. long. Peduncles of the inflorescences mostly single in leaf axils and nearly always unbranched. Male peduncle 2.5–6 cm. long; clusters of male flowers 5–8 mm. in diam. Female peduncle 1–5 cm. long; clusters of female flowers 5–10 mm. in diam.

Malawi. N: Rumphi Distr., on the way to Mwembwe, 9.iv.1981, *Salubeni & Tawakali* 3080 (MAL; SRGH).
Also in Ethiopia, Burundi, Rwanda, Zaire, Uganda and Tanzania.
Evergreen forest, recorded as an epiphyte.

This subspecies represents a comparatively well defined group of populations which can be recognised by the thin, membranous lamina, the usually cuneate leaf bases and the female inflorescences usually consisting of a single cluster of flowers. Typical material of subsp. *rwandensis* has been collected in large quantity from E. Zaire, Rwanda, Burundi, and adjacent parts of Uganda, especially from the montane forests at altitudes between 1800 and 2500 m. Collections of what appears to be this subspecies have also been made from SW. Ethiopia and SW. Tanzania in the area adjacent to Malawi. The description above is based on the entire material of the subspecies. The above specimen is the only record for the Flora Zambesiaca area so far, unfortunately it has rather sparse and immature inflorescences. Its identity is therefore not absolutely certain, however, its overall appearance agrees well with material of typical subsp. *rwandensis*. The presence of the subspecies in SW. Tanzania supports the assumption that this subspecies occurs in the evergreen forests of N. Malawi.

4. **Pilea usambarensis** Engl., Pflanzenw. Ost-Afr. **C**: 163 (1895). —Rendle in F.T.A. **6**, 2: 275 (1917). —Hauman in F.C.B. **1**: 202 (1948). —Friis in Kew Bull. **44**: 582 (1989); in F.T.E.A., Urticaceae: 36 (1989). Type from Tanzania.

Erect or ascending, presumably perennial herbs, monoecious; stems up to 10(50) cm. tall, thin and juicy, prostrate or erect, unbranched or branched, sometimes profusely branched, glabrescent. Leaves of a pair subequal or one with a longer petiole than the other, 1–3(11) × 0.8–2.5(4.5) cm., variable in size and shape from subcircular to ovate, lanceolate or elongate; apex acuminate or rounded; base cuneate, rounded or truncate; margin coarsely dentate or serrate with 4–9(15) teeth on each half of lamina; lamina upper surface with scattered stiff hairs or glabrous, cystoliths linear, lower surface glabrous to puberulous with scattered hydathodes. Petiole (0.5)1–5.5 cm. long, glabrous. Stipules up to 2.5 mm. long, ovate to almost rim-like, with a rounded base. Inflorescences axillary, usually 2 together in upper leaf axils, usually the pair of the same sex; flowers in stalked heads or interrupted spikes with 2–3 glomerules along the axis. Male clusters up to 8(13) mm. in diam., male flowers on pedicels up to 3 mm. long, 3(4)-merous, with apiculate tepals. Female clusters up to 6 mm. in diam., female flowers on pedicels 1–2 mm. long, 3-merous, tepals with appendages as in the male flower, middle tepal lanceolate, up to 1.5 mm. long, lateral tepals shorter. Achene compressed, ± narrowly ovoid, up to 1 mm. long, dull brown.

Malawi. N: South Viphya, Kawandama Forest, 1800 m., female fl. 15.v.1983, *Dowsett-Lemaire* 716 (K).
Also in Kenya and Tanzania. In montane, high rainfall, evergreen forest, often in deep shade. Growing on the forest floor or as an epiphyte on trunks of tree ferns.
Müller 3411 (SRGH) from Zimbabwe (Nyanga Distr., Nyazengu Gorge, 1800 m.) is sterile, but should perhaps also be included here.
In Kenya and Tanzania two additional varieties are recognised; var. *veronicifolia* (Engl.) Friis in which the leaves are subcircular and ± as broad as long and var. *engleri*, recognised by its elliptic or lanceolate leaves which are widest near the middle. The Flora Zambesiaca material belongs to the typical variety and is distinguished by its ovate leaves clearly widest near the base.

6. LECANTHUS Wedd.

Lecanthus Wedd. in Ann. Sci. Nat., Bot. Sér. 4, **1**: 187 (1854).

Annual or short lived perennial, monoecious or apparently dioecious herbs. Cystoliths elongated. Leaves opposite, petiolate. Stipules intrapetiolar, fused. Inflorescences single in the leaf axils, unisexual, rarely bisexual, pedunculate; receptacles surrounded by fused bracts. Male inflorescences usually small, ± bell- or cup-shaped, usually with less than 10 flowers; male flowers 4–5-merous, with a rudimentary ovary. Female inflorescences larger, with many flowers on a flat, disk-shaped receptacle; flowers (3)4-merous, tepals ± markedly cucullate, ovary erect, stigma penicillate, staminodes present. Achene ovoid, verrucose.

This genus consists of one very polymorphic species distributed in the mountains of the Old World tropics.

Lecanthus peduncularis (Royle) Wedd. in DC., Prodr. **16**, 1: 164 (1869). —Rendle in F.T.A. **6**, 2: 276 (1917). —Keay in F.W.T.A., ed. 2, **1**: 621 (1958). —Letouzey in Fl. Cameroun **8**: 180 (1969). Type from Nepal.

Procris peduncularis Royle, Illustr. Bot. Himalaya Mt. **2**: t. 83 (1839). Type as above.
Procris obovata Royle, Illustr. Bot. Himalaya Mt. **2**: t. 83 (1839). Type from Nepal.
Elatostema oppositifolium N.A. Dalz. in Hook. Journ. Bot. **3**: 179 (1851). Type from India.
Elatostema ovatum Wight, Ic. Pl. Ind. Or. **6**: t. 1985 (1853). Type from India.
Lecanthus wightii Wedd. in Ann. Sci. Nat., Bot. Sér. 4, **1**: 187 (1854), nom. illegit. Type as for *Elatostema ovatum.*
Lecanthus major Wedd. in Ann. Sci. Nat., Bot. Sér. 4, **1**: 187 (1854), nom. illegit. Type as for *Elatostema oppositifolium.*
Lecanthus wallichii Wedd. in Ann. Sci. Nat., Bot. Sér. 4, **1**: 187 (1854), nom. nud.

Annual herbs to 10 cm. tall, with juicy stems branched from the lower nodes, hairs short, stiff. Leaves 1–1.5 × 0.5–1 cm., ovate, apex acuminate, base cuneate, margin serrate, almost glabrous above, puberulous below. Petioles up to c. 1 cm. long. Inflorescences on peduncles 0.5–1 cm. long, male inflorescences campanulate, up to 3 mm. in diam., females discoid up to 6 mm. in diam. Male flowers pedicellate, c. 0.5 mm. in diam. Female flowers pedicellate or sessile, up to 0.8 mm. long. Achene up to 1 mm. long.

Although Letouzey, in Fl. Cameroun **8**: 183 (1968), indicated that this species occurs in Malawi no material has been seen at BM, FHO, K, MAL or SRGH.

Recorded from Cameroon, Bioko, E. Zaire, Ethiopia, and tropical Asia from India to S. China, Indonesia, Philippines; perhaps also in Tahiti. At the base of boulders or in rock crevices in moist, shady places, often with moss.

7. PROCRIS Juss.

Procris Juss., Gen. Pl.: 403 (1789).

Perennial herbs, often epiphytic or epilithic, monoecious or dioecious by abortion; stems juicy or succulent. Leaves distichous, opposite and heteromorphous with one leaf of each pair reduced in size, or leaves apparently alternate with one in each pair completely reduced; lamina crenate or serrate, rather fleshy, penninerved, with linear cystoliths. Stipules intrapetiolar, fused. Inflorescences unisexual, in the axils of existing or fallen leaves; male flowers in small pedunculate glomerules; female inflorescences capitate with fleshy receptacles on short fleshy peduncles. Male flowers pedicellate, pentamerous (rarely tetramerous), with a rudimentary pistillode. Female flowers sessile with a persisting perianth of 4–5 free tepals; ovary erect, symmetrical, with a penicillate stigma. Achene ovoid, surrounded at the base by the accrescent perianth.

A genus of about 10 species, most of these distributed in humid habitats in the Old World tropics.

Procris crenata C. B. Robinson in Philipp. Journ. Sci., Bot. **5**: 507 (1911). —Keay in F.W.T.A., ed. 2, **1**: 620 (1958). —Letouzey in Fl. Cameroun **8**: 155 (1968). —Friis in F.T.E.A., Urticaceae: 42 (1989). TAB. **29**. Type from the Philippines.

Procris wightiana sensu Rendle in F.T.A. **6**, 2: 283 (1917). —Schroeter in Fedde Repert. **45**: 191 (1938). —Hauman in F.C.B. **1**: 208 (1948) non Wedd. (1856) nom. illegit.

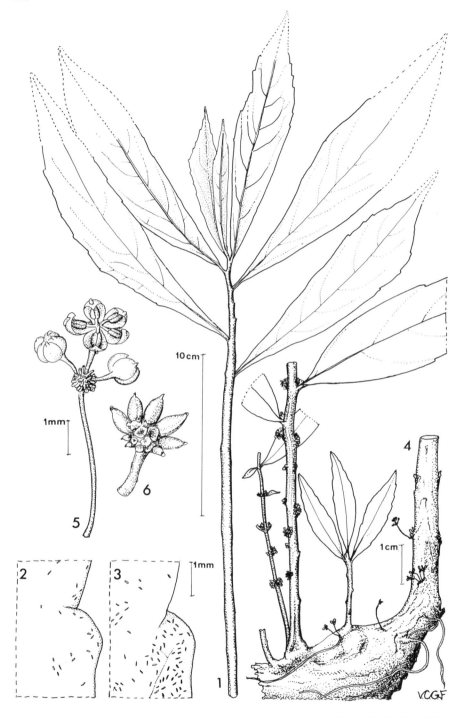

Tab. 29. PROCRIS CRENATA. 1, stem with leaves; 2, leaf, upper surface, detail; 3, leaf, lower surface, detail, 1–3 *Müller* 3319; 4, basal part of stem; 5, male inflorescence, 4–5 *Jacques-Felix* 903; 6, female inflorescence with older flowers, *Banda* 653. Drawn by Victoria C. Friis. From F.T.E.A.

Procris laevigata auct. —Wedd. in DC. Prodr. **16**, 1: 192 (1869), non Blume.

Perennial monoecious herbs, from short, thick, prostrate, ± thickened rhizomes. Stems succulent, usually unbranched, up to c. 40(50) cm. tall, 1–2 cm. in diam. when fresh, shrinking to 4–5 mm. when dry. Leaves opposite and heteromorphous with the reduced leaves of heteromorphous pairs soon falling, usually concentrated in the upper half of the stem, petiolate; lamina up to 12 × 3 cm., somewhat asymmetrical around the midnerve, base asymmetrically cuneate, margin serrate, apex long acuminate, glabrous on both sides, lateral nerves 6–10 on either side of the midnerve. Petioles 2–4 mm. long. Stipules fused, intrapetiolar, 1–3 mm. long. Inflorescences in the axils of current or fallen leaves, pedunculate. Male inflorescences rather rare, usually on the lower part of the plant; peduncles 5–10 mm. long, thin; receptacle small, fleshy, with 3–10 male flowers on 1–3 mm. long pedicels; male flowers (4)5-merous, with corniculate tepals, stamens equalling the tepals in number, ovary rudimentary. Female inflorescences in the upper leaf-axils, on rather stout peduncles, 2–6 mm. long; receptacle fleshy with 20–50 sessile female flowers; female perianth of (4)5 short, broadly ovate tepals, these covering the lower part of the erect, symmetrical, ovoid ovary which is crowned by a penicillate stigma. Achene 1–1.5 mm. long, surrounded at the base by the persistent perianth.

Zimbabwe. E: Penhalonga, Revue River, Manicaland Forest, ster. 24.ix.1950, *Chase* 2901 (BM; LMU). **Malawi**. N: Mzimba Distr., Viphya Plateau, Chikangawa, female fl. 22.iv.1965, *Banda* 653 (SRGH). **Mozambique**. MS: Manica Prov., Penhalonga, near Border Farm, ster. 8.ix.1957, *Chase* 6714 (SRGH).
    Widespread, but apparently rather rare in the humid uplands of tropical Africa, from W. tropical Africa, Cameroon, Congo and Uganda, south through E. Africa to the Flora Zambesiaca area; also in Madagascar, and widespread in tropical Asia, as far east as the Philippines. In upland evergreen forest; 1300–c. 1500 m.

8. ELATOSTEMA J.R. & G. Forster

Elatostema J.R. & G. Forster, Char. Gen. Pl.: 105 (1776) nom. conserv.

Erect annual or perennial, monoecious herbs. Stems prostrate and ascending. Cystoliths linear. Leaves distichous, apparently alternate, sessile or very shortly petiolate, asymmetrical; stipules fused, intrapetiolar, (the presence of 2 pairs of stipules indicates that the genus has opposite leaves but one leaf of each pair is rudimentary or totally reduced). Inflorescences unisexual, axillary, sessile (in the Flora Zambesiaca area), consisting of very densely clustered flowers and surrounded by almost free bracts in the male inflorescences or by somewhat fused bracts in the female inflorescences. Male flowers 4–5-merous, tepals generally with a horn-like dorsal appendage; rudimentary ovary absent or indistinct. Female flowers with 3–5 very reduced tepals and 3 scale-like staminodes, sometimes sterile; ovary ovoid with a penicillate stigma. Achene ovoid, small, ejected by the reflexing of the staminodes.

A genus of about 200 species, often very polymorphic, occurring in the tropics of the Old World. One species has been shown to be apogamous.

The interpretation of the very small and compact inflorescence of *Elatostema* is not yet entirely clear; see Letouzey in Fl. Cameroun 8: 144–145 (1968) for further discussion. The taxonomy of *Elatostema* in tropical Africa is not yet satisfactory.
    A special terminology is practical for describing the two halves of the asymmetrical leaves; the lamina plane agrees with the stem axis and the "proximal half" is the half nearest the stem (usually narrow); the "distal half" is the half away from the stem (usually wider).

1. Leaves with more than (16)25 teeth on the distal margin of the lamina; lamina drying a fresh green, the longest leaves usually more than (15)20 cm. long   -   -   - 1. *welwitschii*
– Leaves with fewer than 25 teeth on each margin; lamina dark green, drying greyish or almost blackish green, the longest leaves usually less than 15 cm. long   -   -   -   - 2
2. Longest stipule (usually that of the reduced leaf) (5)6–9(12) mm. long; number of teeth on distal side of lamina (12)16–22; lamina usually glabrous above; diam. of female inflorescence 10–12 mm.; widest bract of male inflorescence 3–5 mm. wide.   -   -   -   - 2. *paivaeanum*
– Longest stipule (1)2–4(5) mm. long; number of teeth on distal side of lamina 5–10(13); lamina above usually with stiff, scattered hairs; diam. of female inflorescences 4–8 mm.; widest bracts of male inflorescences 1–2.5 mm. wide   -   -   -   -   -   -   -   - 3. *monticola*

1. **Elatostema welwitschii** Engl., Bot. Jahrb. **33**: 124 (1902). —Rendle in F.T.A. **6**, 2: 278 (1917). —Hauman in F.C.B. **1**: 206 (1948). —Robyns in Fl. Parc Nat. Alb. **1**: 82 (1948). —Keay in F.W.T.A. ed. 2, **1**: 620 (1958). —Letouzey in Fl. Cameroun **8**: 151 (1968). —Friis in F.T.E.A., Urticaceae: 39 (1989). TAB. **30**, fig. A. Syntypes from Angola, Cameroon and São Tomé.
    Elatostema henriquesii Engl., Bot. Jahrb. **33**: 125 (1902). Type from São Tomé.
    Elatostema welwitschii var. *cameroonense* Rendle in Journ. Bot. **55**: 201 (1917). Type from Cameroon.

Perennial herbs up to c. 60 cm. tall from creeping rhizomes. Stems erect, fleshy, little-branched, pubescent or hispid. Leaves restricted to the upper half of the stem, shortly petiolate or sessile, lamina asymmetrical, up to 24 × 8 cm., elliptic to ovate; apex long acuminate; base asymmetrical, cuneate on both sides, but the distal half arising about 5 mm. higher up the midnerve than the proximal half; margin finely serrate, with 20–35 teeth on each half of lamina (each tooth terminated by a small tuft of hairs); upper surface of lamina fresh green, with a few stiff hairs, lower surface pale green and with numerous hairs on the nerves, lateral nerves 4–8 on either side. Inflorescences unisexual, often male and female on the same shoot, subsessile or sessile in leaf axils. Male inflorescences consisting of clusters of flowers up to 8 mm. in diam., surrounded by broadly ovate, ciliate, bracts up to 3 × 3 mm.; flowers on pedicels up to 2 mm. long; perianth up to c. 2 mm. in diam., 5-merous, each tepal with a long subapical appendage; stamens the same number as the tepals. Female inflorescences quadrangular or 4-lobed, up to 2.5 cm. in diam., surrounded by triangular bracts up to 1.5 × 1.5 mm.; flowers numerous, pedicellate, perianth extremely reduced consisting of c. 4 segments, staminodes 3 inflexed, ovary c. 0.5 mm. long, with fine longitudinal ridges. Achene up to 3 mm. long, with longitudinal ridges.

Malawi. N: Chitipa Distr., Misuku Hills, Mugesse Forest, 1800 m., male & female fls. 25.iv.1977, *Pawek* (K; MO; SRGH).
    Also in S. Nigeria, Cameroon, Congo, Zaire, Uganda, Tanzania, and Angola. In evergreen rainforest, in forest floor shade; 1400–1800 m.

2. **Elatostema paivaeanum** Wedd. in DC., Prodr. **16**, 1: 178 (1869). —Rendle in F.T.A. **6**, 2: 279 (1917). —Hauman in F.C.B. **1**: 207 (1948). —Keay in F.W.T.A., ed. 2, **1**: 620 (1958). —Letouzey in Fl. Cameroun **8**: 147 (1968). —Friis in F.T.E.A., Urticaceae: 39 (1989). TAB. **30**, fig. B. Type from Bioko.
    Elatostema preussii Engl., Bot. Jahrb. **33**: 126 (1902). Type from Cameroon.
    Elatostema paivaeanum var. *conrauanum* Engl., Bot. Jahrb. **33**: 126 (1902). Type not indicated.
    Elatostema gabonense Schroeter in Fedde Repert. **47**: 217 (1939). Type from Gabon.

Annual or perennial herbs up to c. 40(100) cm. tall from creeping rhizomes. Stems usually unbranched, glabrous or somewhat pubescent. Leaves restricted to the upper half of the stem, subsessile; lamina asymmetrical, up to 15 × 6 cm., obliquely elliptic to ovate; apex acuminate; base oblique (distal half auriculate or subcordate, proximal half cuneate); margin serrate; upper surface of lamina dark green and glabrous or with a few scattered hairs, lower surface paler green and pubescent on the nerves, lateral nerves anastomosing on each side of the midnerve, 5–6 on distal side, 3–4 on proximal side. Stipules linear, 8–12 mm. long. Inflorescences usually unisexual, often male and female on the same shoot, sessile. Male inflorescences up to 1.5 cm. in diam., surrounded by broadly ovate bracts, up to 5 × 5 mm.; flowers on pedicels up to 2 mm. long, perianth up to 2 mm. in diam., 4-merous, each tepal with a long subapical appendix, stamens in the same number as the tepals. Female inflorescences about as large as the male ones, disk-shaped or lobed, surrounded by bracts almost as large as those in the male inflorescences; flowers on pedicels 1–2 mm. long, 3-merous, with a reduced perianth and large staminodes. Achenes shortly stalked, up to 1 mm. long, with many longitudinal ridges.

Malawi. N: Chitipa Distr., Misuku Hills, female fl. 11.i.1959, *Robinson* 3152 (K; PRE; SRGH).
    Also in W. tropical Africa, Cameroon, Gabon, (?)Zaire, and Tanzania. Evergreen rainforest in moist shade of forest floor; 1500–1600 m.

3. **Elatostema monticola** Hook. f. in Journ. Linn. Soc., Bot. **7**: 216 (1864) as "*Elatostemma*". —Wedd. in DC., Prodr. **16**, 1: 187 (1869). —Rendle in F.T.A. **6**, 2: 281 (1917). —Hauman in F.C.B. **1**: 207 (1948). —Keay in F.W.T.A., ed. 2, **1**: 620 (1958). —Lambinon in Bull. Soc. Roy. Bot. Belg. **91**: 210 (1959). —Letouzey in Fl. Cameroun **8**: 141 (1968). —Friis in F.T.E.A., Urticaceae: 41 (1989). TAB. **30**, figs. C1–C6. Type from Cameroon.
    Elatostema orientale Engl., Pflanzenw. Ost-Afr. **C**: 164 (1895). —Rendle in F.T.A. **6**, 2: 280

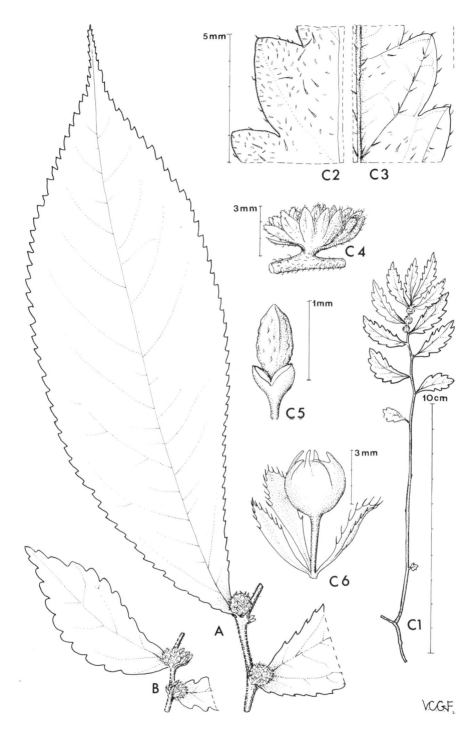

Tab. 30. A.—ELATOSTEMA WELWITSCHII. A, detail of flowering stem with leaf, *Grosvenor & Renz* 1227. B.—ELATOSTEMA PAIVAEANUM. B, detail of flowering stem with leaf, *Grosvenor & Renz* 1228. C.—ELATOSTEMA MONTICOLA. C1, habit; C2, detail of leaf from above; C3, detail of leaf from below; C4, inflorescence with female flowers; C5, old female flower with achene, C1–5 *Chase* 8152; C6, male flower in bud, *Letouzey* 8899. Drawn by Victoria C. Friis. From F.T.E.A.

(1917). —Hauman in F.C.B. **1**: 206 (1948). —Agnew, Upland Kenya Wild Fl.: 323 (1974). Type from Tanzania, destroyed.

    Elatostema kiwuense Engl., Pflanzenw. Afr. **3**, 1: 60 (1915) nom. nud. —Peter in Fedde, Repert., Beih. **40**, 2: 130–131 (1932) sine descr. —Hauman in F.C.B. **1**: 207 (1948) cum descr. gall.

    Elatostema orientale var. *longiacuminatum* De Wild., Pl. Bequaert. **5**: 383 (1932) as "*longeacuminata*". Type from E. Zaire.

    Elatostema longiacuminatum (De Wild.) Hauman in F.C.B. **1**: 207 (1948) —Lambinon in Bull. Soc. Roy. Bot. Belg. **91**: 211 (1959). Type as above.

Annual or perennial herbs, apparently dioecious, rarely monoecious. Stems erect or ascending, to c. 30 cm. tall from prostrate, ± branching rhizomes, often with short crisp hairs in the upper part of the stem. Leaves restricted to the upper part of the erect stems, sessile or subsessile; lamina usually asymmetrical and curving towards top of stem, up to 8 × 2 cm., obovate to lanceolate; apex obtuse, acute or long acuminate; base oblique, proximal side crenate, distal side auriculate or subcordate, cuneate; margin serrate, distal margin with 5–10 teeth, proximal margin with fewer teeth; upper surface of lamina glabrescent, cystoliths linear, numerous, lower surface puberulous mainly on the nerves. Stipules fused to about the middle, up to 4 mm. long, lanceolate. Inflorescences sessile, unisexual, axillary, sometimes contiguous toward the stem apex. Male inflorescences infrequent, mainly in the axils of the lower leaves, up to 3(8) mm. in diam.; bracts lanceolate, ciliate; flowers on pedicels c. 2 mm. long, perianth 4-merous, tepals with dorsal horn-like appendages. Female inflorescences frequent, in the axils of the upper leaves, bell- or disk-shaped, up to 5 mm. in diam., surrounded by bracts; flowers almost sessile, but probably becoming pedicellate during ripening of the achene, tepals 3 very reduced, staminodes 3, ovary erect, stigma penicillate. Sterile pedicellate flowers apparently present, but possibly these are fertile flowers from which the achenes have been ejected. Achene c. 1 mm. long, longitudinally ridged.

    Zambia. E: Nyika Plateau, Chowo Forest, 2000 m., male & female fls. 14.iv.1991, *Friis* 6324 (C; K; MAL). **Zimbabwe**. E: Mutare Distr., Vumba Mts., Nyamataka River, female fl. 29.iii.1964, *Chase* 8152 (K; PRE; SRGH). **Mozambique**. MS: Gorongosa, Mt. Nhandore, 1200 m., female fl. 6.v.1964, *Torre & Paiva* 12281 (LISC).

    Also in S. Nigeria, Cameroon, Bioko, E. Zaire, Rwanda, S. Sudan (Imatong Mts.), Ethiopia, Uganda, Kenya, and Tanzania. Evergreen montane forest in moist shade of forest floor, sometimes in shade in moist rocky places; 1200–2200 m. This species would also be expected to occur in Malawi.

    I have followed Lambinon (Bull. Soc. Bot. Belg. **91**: 209–211 (1959)) and consider *E. orientale* and *E. kiwuense* as synonyms of *E. monticola*. *E. kiwuense* seems to represent the small annual forms which may deserve recognition at the infraspecific level, but not enough material is yet available on which to make a decision. Unlike Lambinon, I am unable to distinguish *E. monticola* from *E. longiacuminatum* in E. Africa, and have therefore tentatively placed the latter in synonymy of the former. The material from the Flora Zambesiaca area is unusually homogeneous in comparison with that from E. Africa, and differs somewhat in having smaller and less asymmetrical leaves. I have treated these Flora Zambesiaca populations as *E. monticola* in the wide sense, as I am at the moment unable to suggest a better taxonomic position or status for these plants.

9. BOEHMERIA Jacq.

Boehmeria Jacq., Enum. Pl. Carib.: 9, 31 (1760).

    Shrubs or small trees, sometimes woody-based herbaceous perennials, monoecious or dioecious, usually ± pubescent, stinging hairs absent. Cystoliths dot-like. Leaves alternate or opposite, petiolate, lamina triplinerved, equal- or unequal-sided, dentate sometimes lobed; stipules lateral, free or ± connate. Inflorescences axillary, usually spike-like, with the flowers gathered in unisexual glomerules along the inflorescence axis (rarely sessile and globular); bracts small, deciduous. Male flowers (3)4(5)-merous with membranaceous tepals, rudimentary ovaries present or absent. Female flowers with tubular perianths, constricted and 2–4-toothed at the apices, completely enclosing the ovaries. Style filiform, not articulated, with a one-sided filiform stigma. Achene completely enclosed in and partly adnate to the persistent perianth.

    A genus of about 50 species, pantropical and in warm temperate regions.

    Boehmeria nivea (L.) Gaudich. is widely cultivated in the tropics for its fibre; it differs from the native species in having alternate leaves which are white tomentose beneath. It has been recorded from Mozambique (Mangorre, Estação Experimental da Malamba, female fl. 12.iv.1954, *Barbosa &*

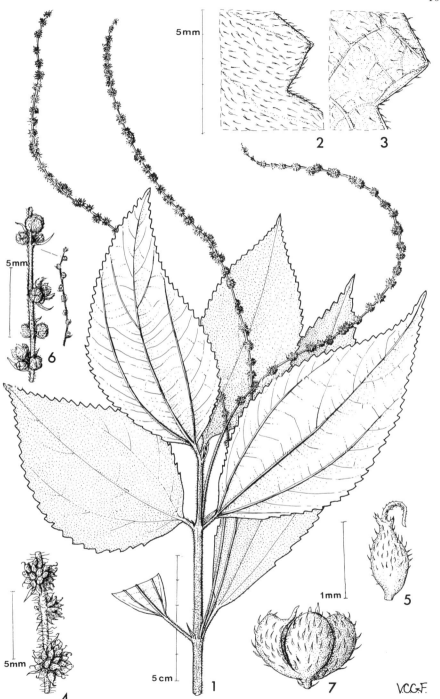

Tab. 31. BOEHMERIA MACROPHYLLA. 1, flowering stem; 2, leaf, upper surface, detail; 3, leaf, lower surface, detail; 4, part of female inflorescence; 5, old female flower with achene enclosed in persisting perianth, 1–5 *Brummitt, Chisumpa & Polhill* 13906; 6, part of male inflorescence; 7, male flower, bud, 6–7 *Simon & Williamson* 1923. Drawn by Victoria C. Friis. From F.T.E.A.

Balsinhas 5521 (LISC)). It is possible that it has been introduced elsewhere in the Flora Zambesiaca area.

Boehmeria macrophylla Hornem., Hort. Reg. Bot. Hafn. **2**: 890 (1815). —Friis & Marais in Kew Bull. **37**: 164 (1982). —Marais in Fl. Masc. 161, Urtic.: 28 (1985). —Friis in F.T.E.A., Urticaceae: 44 (1989). TAB. **31**. Type a plant of unknown origin cultivated in the Copenhagen Botanical Garden.

> *Boehmeria platyphylla* D. Don, Prodr. Fl. Nepal.: 60 (1825). —Rendle in F.T.A. **6**, 2: 285 (1917). —Hauman in F.C.B. **1**: 210 (1948). —Keay in F.W.T.A., ed. 2, **1**: 622 (1958). —Letouzey in Fl. Cameroun **8**: 196 (1968). —Agnew, Upland Kenya Wild Fl.: 325 (1974). Type from Nepal.
>
> *Boehmeria platyphylla* var. *nigeriana* Wedd. in DC., Prodr. **16**, 1: 213 (1869). —Rendle in F.T.A. **6**, 2: 286 (1917). —Hauman in F.C.B. **1**: 211 (1948). Type from Nigeria.
>
> *Boehmeria platyphylla* var. *tomentosa* Wedd. in DC., Prodr. **16**, 1: 212 (1869). —Hauman in F.C.B. **1**: 211 (1948). —Léandri in Fl. Madag. **56**, Urtic.: 75 (1965). Types from India, Java and Madagascar.
>
> *Boehmeria platyphylla* var. *ugandensis* Rendle in Journ. Bot. **55**: 202 (1917); in F.T.A. **6**, 2: 286 (1917). Type from Uganda.
>
> *Boehmeria platyphylla* var. *angolensis* Rendle in Journ. Bot. **55**: 201 (1917); in F.T.A. **6**, 2: 286 (1917). —Hauman in F.C.B. **1**: 211 (1948). Type from Angola.
>
> *Boehmeria platyphylla* forma *platyphylla* (as forma "*typica*") —Hauman in F.C.B. **1**: 211 (1948). Type as for *B. platyphylla*.

Erect, shrubby or herbaceous perennial plants, up to 3 m. tall; stems ± thick, sparsely branched, or with long, often overhanging, thinner branches, younger stems pubescent, older ones glabrescent. Leaves opposite, sometimes the pairs heteromorphic, 15–30 × 7–15 cm., ovate, sometimes very broadly so, to lanceolate; apex acuminate sometimes almost caudate; base cordate, rounded or cuneate; margin serrate with 15–35 teeth on each side; upper surface of lamina indumentum of unequally long, rather stiff hairs, or glabrescent, lower surface indumentum of softer hairs more uniform in length. Petiole 2.5–15 cm. long; stipules lateral, membranous, brown, to c. 8 mm. long, lanceolate. Inflorescences unisexual, single in upper leaf axils, interrupted pendent spikes up to 30(50) cm. long, with the flowers in sessile dense glomerules; each glomerule subtended by a caducous bract; male glomerules 3–5 mm. in diam.; female glomerules 5–8 mm. in diam., young glomerules covered by bracts. Male flowers sessile, perianth up to 2 mm. long, tetramerous, tepals with a ± pronounced dorsal gibbosity. Female flowers sessile, perianth up to 3 mm. long, bidentate at apex, pubescent in the upper part. Achenes up to 3 mm. long, enclosed in the persistent perianth.

Zambia. N: Samfya, male fl. 5.v.1958, *Fanshawe* 4404 (K; SRGH). W: Luakera Forest, 14 km. N. of Mwinilunga, male & female fls. 20.i.1975, *Brummitt, Chisumpa & Polhill* 13906 (K; SRGH). **Zimbabwe.** E: Chimanimani Distr., Rusitu Mission, Ngorima Reserve, 1300 m., male & female fls. 13.i.1965, *Chase* 8226 (K; PRE; SRGH). **Malawi.** N: Chitipa Distr., Misuku Hills, Mugesse Forest, female fl. 15.iii.1977, *Grosvenor & Renz* 1231 (K; PRE; SRGH). C: Ntchisi Mt., 1450 m., male & female fls. 19.ii.1959, *Robson* 1660 (K; PRE; SRGH). S: Zomba Plateau, 1400 m., female fl., fr. 28.v.1946, *Brass* 16057 (K; PRE; SRGH). **Mozambique.** N: Lichinga (Vila Cabral), Serra de Massangulo, 1450 m., female fl. 15.iii.1964, *Torre & Paiva* 11044 (LISC). Z: Zambezia Prov., Serra Morrumbala, 1080 m., male fl. 13.xii.1971, *Müller & Pope* 2027 (K; SRGH). T: Angónia, Rio Dzenza, 1400 m., female fl., 10.xii.1980, *Stefanesco & Nyongani* 577 (SRGH). MS: Mavita, female fl. 10.iv.1948, *Barbosa* 1420 (LISC).

Widespread in tropical Africa, from Guinean Republic east to Ethiopia and south to Angola; also in tropical Asia and Indonesia. Understorey plant of mixed evergreen forest, rainforest, riverine forests and swamp forests (mushitu), often in dense tangles beside streams, in moist, rocky places; 300–1300(1500) m.

Letouzey, in Fl. Cameroun **8**: 196–207 (1968), subdivided this species into five forms in the Cameroons, not formally named, but designated by the letters A–E. I have felt unable to subdivide the Flora Zambesiaca material in a similar manner.

10. POUZOLZIA Gaudich.

Pouzolzia Gaudich. in Freyc., Voy. Monde, Bot.: 503 (1830). —Friis & Jellis in Kew Bull. **39**: 587–601 (1984).

Erect annual or perennial herbs, or shrubs, monoecious. Cystoliths dot-like. Leaves alternate, opposite or verticillate, petiolate or sessile, entire or dentate; stipules free, lateral. Inflorescences of compact, axillary bisexual glomerules. Male flowers pedicellate,

4–5-merous, tepals rarely with transverse crests or wings, ovaries rudimentary or absent. Female flowers sessile, with an indefinite number of completely fused tepals which enclose the erect ovaries; staminodes absent; stigma sessile, filiform with lateral papillae, deciduous. Achene ± compressed ovoid, completely enclosed in the persistent membranaceous perianth.

A genus of about 50 species, mainly in the Old World tropics, occurring in both dry and humid habitats.

1. Margin of fully developed leaves entire - - - - - - - - - 2
- Margin of fully developed leaves serrate - - - - - - - - 3
2. Leaf lower surface with a white tomentum - - - - - - - 1. *mixta*
- Leaf lower surface without a white tomentum - - - - - - -2. *shirensis*
3. Shrubs or perennial herbs to 3 m. tall; inflorescence bracts linear, about as long as or shorter than the flowers - - - - - - - - - - - 3. *parasitica*
- Low annual herbs; inflorescence bracts broad, lanceolate, about as long as the inflorescence itself - - - - - - - - - - - - - 4. *bracteosa*

1. **Pouzolzia mixta** Solms-Laub. in Sitz. Ges. Nat. Freunde Berlin **1864**: 1 (1864); in Schweinf., Beitr. Fl. Aethiop.: 188 (1867). —Schweinf. in Bull. Herb. Boiss. **4**, Appendix 2: 146 (1896). —Rendle in F.T.A. **6**, 2: 291 (1917). —F.W. Andr., Fl. Pl. Anglo-Egypt. Sudan **2**: 279 (1952). —Friis & Jellis in Kew Bull. **39**: 590 (1984). —Friis in F.T.E.A., Urticaceae: 48 (1989). TAB. **32**, figs. A1–A6. Type from the border between Sudan and Ethiopia.
    Pouzolzia hypoleuca Wedd. in DC., Prodr. **16**, 1: 227 (1869). —Rendle in F.T.A. **6**, 2: 291 (1917). —N.E. Br. in F.C. **5**, 2: 551 (1925). —Roessler in Merxm., Prodr. Fl. SW. Afr. **17**: 6 (1967). Type: Mozambique, Moramballa, *Kirk* s.n. (holotype K).
    Pouzolzia arabica Defl., Voy. Yemen: 206 (1889). Type from Yemen.
    Pouzolzia huillensis Hiern, Cat. Afr. Pl. Welw. **1**, 4: 993 (1900). —Rendle in F.T.A. **6**, 2: 290 (1917). Type from Angola.
    Pouzolzia fruticosa Engl., Bot. Jahrb. **33**: 127 (1902). —Rendle in F.T.A. **6**, 2: 292 (1917). Type from Ethiopia.

Shrubs to 3(5) m. tall. Stems of soft wood with a spongy pith or hollow centre, branchlets thick, more than 1.5 mm. in diam.; bark longitudinally striate, greyish or reddish-brown, leaf and inflorescence scars prominent; indumentum on young parts white-arachnoid, later glabrescent. Leaves deciduous, 2–6(10) × 1–3(7) cm., ovate; apex acuminate; base obliquely cuneate or rounded; margin entire; upper surface of lamina ± scabridulous with scattered stiff hairs, lower surface with a ± densely white arachnoid tomentum. Petiole 0.8–1.5(2) cm. long. Stipules up to 5 mm. long, lanceolate, puberulous, ciliate. Inflorescences bisexual, in dense axillary clusters, appearing with young leaves, sessile in axils of current leaves or scattered along twigs in axils of fallen leaves; bracts short, linear. Male flowers numerous, on pedicels c. 1 mm. long; perianth c. 1 mm. in diam., (4)5-merous. Female flowers few, sessile; perianth 1.5–2 mm. long, ovoid, fused almost to the apex and completely enclosing the ovary, with 3–8 longitudinal ridges of which 2 are usually larger than the others; stigma 2–4 mm. long, filiform, exserted. Achene c. 2 mm. long, brownish, smooth, enclosed in the persistent perianth.

Botswana. N: c. 20 km. S. of the Botswana border, c. 1500 m., male & female fls. 13.i.1960, *Leach & Noel* 17 (K; PRE; SRGH). SW: N'gami, ster. 5.v.1930, *van Son* in Transvaal Mus. 29033 (PRE). SE: Gaborone, Aedume Park, 1050 m., male & female fls. 22.x.1977, *O.J. Hansen* 3239 (C; GAB; K; PRE; SRGH; UPS; WAG). **Zambia**. N: Mbala Distr., Lake Tanganyika, Crocodile Isl., 780 m., male & female fls. 9.ii.1964, *Richards* 18991 (K; SRGH). C: Lusaka, male & female fls. 13.xii.1968, *Fanshawe* 10489 (K; NDO; SRGH). E: Chadiza, 850 m., male & female fls. 25.xi.1958, *Robson* 695 (K; PRE; SRGH). S: Livingstone Distr., Natebe, 1900 m., male & female fls. 1.xii.1955, *Gilges* 483 (K; PRE; SRGH). **Zimbabwe**. N: Hurungwe Distr., Musukwi (Msukwe) River, Kanyonga Stream, 1000 m., male & female fls. 17.xi.1953, *Wild* 4181 (K; PRE; SRGH; LMU). W: Matobo Distr., Matopos, male & female fls. xii.1959, *Armitage* 312/59 (SRGH). C: Harare Distr., Manyame (Hunyani) Bridge, Prince Edward Dam wall, male & female fls. 22.xi.1960, *Rutherford-Smith* 351 (K; M; PRE; SRGH). E: Mutare Distr., Honde Valley, male & female fls. 5.xi.1948, *Chase* 1343 (K; SRGH). S: Masvingo Distr., Mushandike Nat. Park., male & female fls. 28.ii.1975, *Bezuidenhout* 231 (K; SRGH). **Malawi**. N: Nkhata Bay Distr., Sanga, c. 24 km. S. of the Mzuzu-Nkhata Bay road, 500 m., male & female fls. 17.xii.1972, *Pawek* 6099 (K; MO; SRGH; MAL). C: Lilongwe Distr., Bunda Mt., c. 20 km. S. of Lilongwe, 1250 m., male & female fls. 20.ii.1970, *Brummitt* 8672 (K; SRGH). S: Zomba Distr., Old Naisi road, male & female fls. 22.xi.1979, *Salubeni & Tawakali* 2659 (K; SRGH; MAL). **Mozambique**. N: Niassa Prov., Lichinga (Vila Cabral), Meponda, male & female fls. 9.iv.1961, *M.F. Carvalho* 491 (K; LMU). Z: Milange, Sabelua-Liciro, Mt. Sonelafuti, 600 m., male & female fls., fr. 14.ii.1972, *Correia & Marques* 2642 (LMU). T: Tete, Chidanzua (Chinhunque e Tanara), road between Estima and

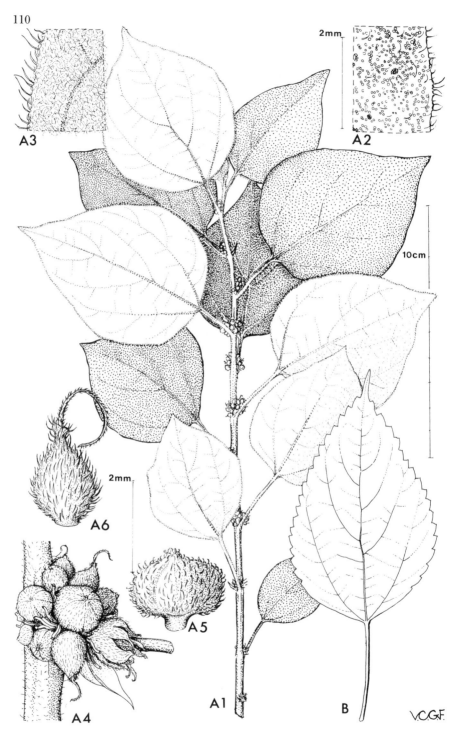

Tab. 32. A.—POUZOLZIA MIXTA. A1, flowering branch; A2, leaf, superior surface, detail; A3,
    leaf, inferior surface, detail; A4, inflorescence, in leaf axil with stipule, male & female flowers;
    A5, bud of male flower; A6, female flower, A1–6 *Norrgrann* 73. B.—POUZOLZIA PARASITICA.
    B, leaf, *Salubeni & Masiye* 2391. Drawn by Victoria C. Friis. From F.T.E.A.

Songo, male & female fls. 4.ii.1972, *Macêdo* 4767 (LISC; LMU; SRGH). MS: Chimoio, Serra de Chindaza, male & female fls. 14.iii.1948, *Garcia* 610 (LISC). M: Sabiè, male & female fls. 30.xi.1944, *Mendonça* 3173 (LISC).

Also in Yemen, E. Sudan, Ethiopia, Tanzania, Angola, Namibia and S. Africa (Transvaal, Natal). In open deciduous woodlands, wooded ravines and riverine thickets usually on rocky escarpments, rocky hillsides, or rocky outcrops, sometimes on sand; 250–1900 m.

This is the most widespread and common species of the family in the Flora Zambesiaca area, where it occurs almost throughout the drier parts. There is a peculiar disjunction in the northern part of the distribution of this species; it occurs in the Yemen, Ethiopia, and E. Sudan, seems to be absent from Kenya and Uganda, and reappears in Tanzania and southern tropical Africa. It has not been possible to suggest a meaningful taxonomic distinction between specimens from these two disjunct areas of distribution.

2. **Pouzolzia shirensis** Rendle in Journ. Bot. **55**: 202 (1917); in F.T.A. **6**, 2: 292 (1917). —Friis & Jellis in Kew Bull. **39**: 590 (1984). Type: Malawi, Shire Highlands, *Scott Elliot* 8679 (BM, lectotype, selected here; K, isolectotype).

Very similar to *P. mixta*. It differs chiefly in the leaves which are green on the lower surface and almost totally devoid of the silvery white arachnoid tomentum which is so characteristic for *P. mixta*.

Malawi. S: Blantyre Distr., Mpatamanga Gorge, on Shire River, 230 m., male & female fls. 9.ii.1970, *Brummitt* 8436 (K).

This taxon is endemic in Malawi. More material is necessary before it can be established whether or not it is distinct. It could very well represent a glabrous form of *P. mixta*.

3. **Pouzolzia parasitica** (Forssk.) Schweinf. in Bull. Herb. Boiss. **4**, Appendix 2: 145 (1896). —Rendle in F.T.A. **6**, 2: 293 (1917). —Hauman in F.C.B. **1**: 215 (1948). —Robyns in Fl. Parc Nat. Alb. **1**: 84 (1948). —Keay in F.W.T.A., ed. 2, **1**: 763 (1958). —Letouzey in Fl. Cameroun **8**: 194 (1968). —Friis & Jellis in Kew Bull. **39**: 593 (1984). —Friis in F.T.E.A., Urticaceae: 51 (1989). TAB. **32**, fig. B. Type from Yemen.
  Urtica parasitica Forssk., Fl. Aegypt.-Arab.: CXXI, 160 (1775). Type as above.
  Urtica muralis Vahl, Symb. Bot. **1**: 77 (1790) nom. illegit. Type as for *Urtica parasitica* Forssk.
  Margarocarpus procridioides Wedd. in Ann. Sci. Nat., Bot. Sér. 4, **1**: 204 (1854). Type from S. Africa (Natal).
  Pouzolzia procridioides (Wedd.) Wedd. in Archiv. Mus. Nat. Hist. Nat., Paris 9 (Monogr. Urtic.): 412 (1856); in DC., Prodr. **16**, 1: 231 (1839). —N.E. Br. in F.C. **5**, 2: 551 (1925). Type as above.
  Boehmeria procridioides (Wedd.) Blume in Mus. Bot. Lugd.-Bat. **2**: 204 (1856). Type as above.

Perennial herb from a woody rhizome. Stems erect or ascending sometimes scrambling, up to c. 1 m. tall, slightly woody at the base, up to c. 5 cm. in diam., covered by a dense indumentum of short, erect hairs, bark greyish brown; ± prostrate stems often rooting at the nodes. Leaves alternate, 4–8(10) × 3–5 cm., ovate to lanceolate; apex acuminate; base cuneate, rounded, or subcordate; margin coarsely serrate; upper surface of lamina with scattered, stiff hairs and punctiform cystoliths, lower surface glabrescent with softer hairs on the nerves. Petiole pubescent, up to 7 cm. long. Stipules lanceolate, up to 7 mm. long. Inflorescences of densely clustered flowers in the axils of the upper leaves, bracts narrowly triangular. Male flowers numerous, pedicels c. 1.5 mm. long; perianth globose, c. 1 mm. in diam., 3–4-merous, vestigial ovary present. Female flowers fewer, sessile; perianth c. 2 mm. long, with 10 faint longitudinal lines; ovary enclosed, stigma erect, 2–4 mm. long, filiform, exserted. Achene c. 2.5 mm. long, shiny, white, enclosed in the persistent perianth.

Zambia. N: Kasama Distr., Chozi Flats, 1200 m., male & female fls. 14.xii.1964, *Richards* 19375 (K; SRGH). W: Ndola, male & female fls. 14.xii.1953, *Fanshawe* 580 (K; SRGH). **Zimbabwe**. C: Wedza Distr., Wedza Mt., ster. 22.v.1968, *Rushworth* 1088 (SRGH). E: Mutare, Ranger Hill, 1200 m., male & female fls. 17.ii.1960, *Chase* 7275 (K; PRE; SRGH). S: Mberengwa Distr., Mt. Buhwa, 1100 m., male & female fls. 1.v.1973, *Biegel, Pope & Simon* 4245 (K; PRE; SRGH). **Malawi**. N: Chitipa Distr., Misuku Hills, Mugesse For. Res., 24.v.1989, *Radcl.-Smith, Pope & Goyder* 5939 (K). C: Dedza Mt., male & female fls. 21.ii.1968, *Salubeni* 973 (K; SRGH). S: Zomba Plateau, male & female fls. 18.xii.1978, *Salubeni & Masiye* 2391 (SRGH). **Mozambique**. MS: Manica Distr., Border Farm, male & female fls., fr. 19.iii.1961, *Chase* 7442 (K; SRGH).

Also in Yemen, Ethiopia, Somalia, Cameroon, E. Zaire, and E. Africa, Angola and S. Africa (Transvaal, Natal); also recorded from tropical S. America. High rainfall deciduous woodland, submontane evergreen forests, gully forests and riverine forest, moist tall grassland; usually in shade of dense vegetation, often on termite hills; 1000–1500 m.

The specimens from N. Zambia differ somewhat from the rest of the material of this species in having several rather thick, erect stems from a large rootstock. This form seems to occur in *Brachystegia* woodland, but is not sufficiently distinct to merit formal taxonomic recognition. Similar erect woodland forms occur in Ethiopia and Somalia.

4. **Pouzolzia bracteosa** Friis in Bol. Soc. Brot. ser. 2, **58**: 202 (1985). Type: Zambia, Luangwa Valley, at Point Bar, 21.iii.1972, *Abel* 490 (SRGH, holotype).

Erect annual herbs up to 45 cm. tall; young branches and petioles with an indumentum of scattered, stiff hairs. Leaves opposite (at least the lower 2 pairs) or alternate, sessile (on short side shoots, grading into bracts), or petiolate; mature leaves up to 6 × 3 cm., lanceolate to ovate; apex acute to acuminate; base cuneate to subtruncate; margin serrate, with 3(5) rounded teeth; lamina upper surface with scattered stiff hairs, lower surface with scattered stiff hairs on the nerves. Petiole up to 3.5 cm. long; stipules up to 3.5 mm. long, linear-lanceolate, with a narrow ciliate apex. Inflorescences bisexual, on shortened axillary branches; bracts numerous, subsessile, intergrading with the leaves, up to 10 × 7 mm., elliptic-ovate, tapering to a c. 3 mm. long petiole-like base or sessile, margins entire, apex acute to rounded, lamina with scattered stiff hairs on both sides. Male flowers apparently few, pedicels up to 1 mm. long, bracteolate, perianth c. 1 mm. in diam., 4-merous, with hooked hairs on the outside of the tepals. Female flowers sessile, perianth completely fused, with 5(6) obscurely marked ridges and scattered stiff hairs on the outside. Stigma not observed. Achene dark brown, glossy, enclosed in the persistent perianth.

Zambia. C: Luangwa Valley, male & female fls. 9.v.1972, *Abel* 640 (SRGH). So far known only from the specimens cited above. Growing on alluvium near river.
Probably related to *Pouzolzia fadenii* Friis & Jellis in Kew Bull. **39**: 593 (1984) from the coast of Kenya, but more material of both taxa is needed before their status can be ascertained.

11. DROGUETIA Gaudich.

Droguetia Gaudich. in Freyc., Voy. Monde, Bot.: 505 (1830). —Friis & Wilmot-Dear in Nordic Journ. Bot. **8**: 36 (1988).
Didymogyne Wedd. in Ann. Sci. Nat., sér. 4, **1**: 207 (1854).
Droguetia sect. *Didymogyne* (Wedd.) Benth. in Gen. Pl. **3**: 394 (1880).

Slender annual or perennial herbs, occasionally subshrubs, with erect or prostrate stems, monoecious or apparently dioecious (by abortion of all bisexual inflorescences leaving only female flowers). Cystoliths dot-like. Leaves alternate or opposite, petiolate, serrate; stipules lateral, free. Inflorescences of sessile clusters in the leaf axils; clusters either bisexual with few–many flowers surrounded by a common campanulate involucrum of fused bracts or inflorescence clusters of many very small involucres each with 1(2) female flowers (the involucres then appear as perianths). Bisexual involucres usually with many male flowers and 1–few female flowers (occasionally all the flowers are male); bracts and flowers with long woolly hairs. Male flowers pedicellate, club-shaped; perianth 3-lobed, one lobe apiculate and provided with prominent bristles; stamen 1, inflexed in bud, later reflexed; rudimentary ovary absent. Female flowers naked, pedicellate, consisting of an ovary covered with a woolly tomentum; stigma sessile, filiform; staminodes absent. Entirely-female involucres are common on certain individuals, and then much smaller and with much fewer flowers than the bisexual involucres. Achene enclosed in the persistent involucre, brown, shiny, often crowned by the persistent stigma.

A genus of 7 species distributed in tropical and S. Africa, Yemen, India, and Java; also on Madagascar and the Mascarenes.

Droguetia iners (Forssk.) Schweinf. in Bull. Herb. Boiss. **4**, Appendix 2: 146 (1896). —Rendle in F.T.A. **6**, 2: 303 (1917). —Hauman in F.C.B. **1**: 217 (1948). —F. W. Andr., Fl. Pl. Anglo-Egypt. Sudan **2**: 276 (1952). —Keay in F.W.T.A., ed. 2, **1**: 622 (1958). —Letouzey in Fl. Cameroun **8**: 213 (1968). —Agnew, Upland Kenya Wild Fl.: 325 (1974). —Friis in F.T.E.A., Urticaceae: 56 (1989). TAB. **33**. Type from Yemen.
    Urtica iners Forssk., Fl. Aegypt.-Arab.: 160 (1775). Type as above.
    Urtica verticillata Vahl, Symb. Bot. **1**: 76 (1790) nom. illegit. Type as for *Urtica iners* Forssk.
    Parietaria capensis Thunb., Prodr. Pl. Cap. **1**: 31 (1794). Type from S. Africa.

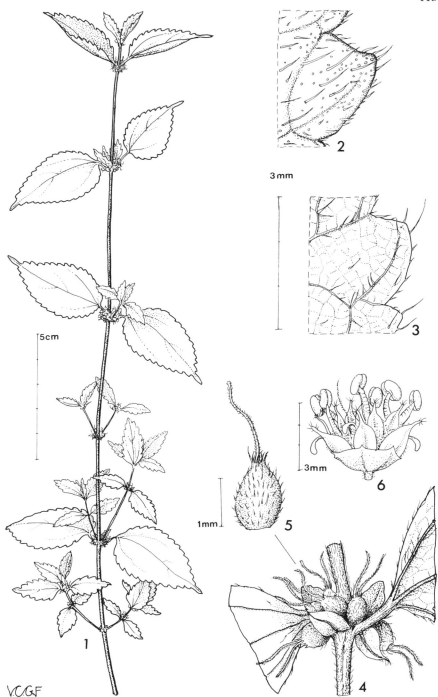

Tab. 33. DROGUETIA INERS. 1, part of flowering stem; 2, leaf, upper surface, detail; 3, leaf, lower surface, detail; 4, compound inflorescence of involucres, each with one or two female flowers; 5, involucre (female inflorescence) with one flower, 1–5 *Brass* 17814; 6, bisexual involucre with one female and several male flowers, *Greenway & Kanuri* 13882. Drawn by Victoria C. Friis.

Boehmeria capensis (Thunb.) Spreng., Syst. Veg., ed. 16, **3**: 844 (1826).
Urtica pauciflora Steud. in Flora **33**: 258 (1850). Type from Ethiopia.
Pouzolzia pauciflora (Steud.) A. Rich., Tent. Fl. Abyss. **2**: 259 (1851). Type as above.
Didymogyne abyssinica Wedd. in Ann. Sci. Nat., Bot. Sér. 4, **1**: 207 (1854) nom. illegit. Type as for
  Urtica pauciflora Steud.
Droguetia diffusa Wedd. in Ann. Sci. Nat., Bot. Sér. 4, **1**: 211 (1854). Type from Ethiopia.
Boehmeria parvifolia Wedd. in Ann. Sci. Nat., Bot. Sér. 4, **1**: 203 (1854). Type from Java.
Boehmeria pauciflora (Steud.) Blume, Mus. Bot. Lugd.-Bat. **2**: 201 (1856). Type as above.
Droguetia pauciflora (Steud.) Wedd. in DC., Prodr. **16**, 1: 235, 58 (1869). Type as above.
Urtica urens var. *iners* (Forssk.) Wedd. in DC., Prodr. **16**, 1: 40 (1869). Type as above.
Droguetia woodii N.E. Br. in F.C. **5**, 2: 561 (1925). Type from S. Africa.
Droguetia thunbergii N.E. Br. in F.C. **5**, 2: 558 (1925) nom. illegit., based on the same type as
  Parietaria capensis Thunb.

Perennial herbs with erect branched stems to c. 1.5 m. tall, and creeping or scrambling stems, rooting at the nodes. Younger stems pubescent or slightly hispid, later glabrescent. Leaves opposite, 3–7 × 2–4 cm., ovate; apex long acuminate; base cuneate or rounded; margin serrate; upper surface of lamina slightly hispid, lower surface with soft hairs on the nerves. Petiole pubescent, 0.4–3 cm. long; stipules up to 0.5 cm. long, lanceolate, mostly persisting. Inflorescences variable, typically consisting of two bisexual involucres (containing male and female flowers) and a reduced branch with few–many female involucres (each containing 1(2) female flowers) in each leaf axil, but any of these elements may be missing. Both types of inflorescence may be found on different plants or on the same plant. Bisexual involucres 4–6 mm. in diam., mostly with many male flowers and a few naked female flowers; male flowers up to c. 2 mm. long. Female involucres often many together, perianth-like, each usually with 1(2) female flowers, up to 1.5 mm. long. Ovary erect, 1–1.5 mm. long; stigma up to c. 3 mm. long. Achene compressed, brownish, c. 1.5 mm. long, enclosed in the slightly enlarged, persistent involucre.

Zimbabwe. E: Mutare Distr., Engwa, 1830 m., female fl. 2.ii.1955, *Exell, Mendonça & Wild* 130 (SRGH). **Malawi**. C: Ntchisi Forest, female fl. 11.vii.1960, *Chapman* 835 (K; SRGH). S: Thyolo (Cholo) Mt., 1350 m., female fl. 26.ix.1946, *Brass* 17814 (K; PRE; SRGH).
  Also in Cameroon, Bioko, E. Zaire, Rwanda, Burundi, S. Sudan (Imatong Mts.), Ethiopia, Uganda, Kenya, Tanzania, Angola and S. Africa (Natal, Cape Province); and in the Yemen. Montane evergreen forest, in ravines or in plantations in forest floor shade; 1350–1850 m.

  Subsp. *burchellii* (N.E. Br.) Friis & Wilmot-Dear differs in the woolly indumentum on the stems and petioles and occurs in the Cape Province of S. Africa where it is ecologically separated from the typical subspecies. Subsp. *urticoides* (Wight) Friis & Wilmot-Dear differs in the woolly indumentum on the inside of the involucre and occurs in S. India and on Java.

12. DIDYMODOXA Wedd.

Didymodoxa Wedd., in Archiv. Mus. Nat. Hist. Nat., Paris 9 (Monogr. Urtic.): 547 (1856); in
DC., Prodr. **16**, 1: 235, 61 (1869). —Friis in Bol. Soc. Brot. **58**: 209–210 (1985). —Friis &
Wilmot-Dear in Nordic Journ. Bot. **8**: 45 (1988).
Australina sensu auct., non Gaudich. (1830).

Slender annual herbs, monoecious or dioecious by abortion. Cystoliths dot-like. Leaves alternate, petiolate, serrate, dentate or entire; stipules lateral, free. Inflorescences of dense sessile clusters, axillary or on reduced branches, bisexual or unisexual by abortion, always without a common involucre, but often with ciliate or bristle-bearing bracts. Male flowers few together, sessile; perianth boat-shaped; stamen single, inflexed; rudimentary ovary absent. Female flowers sessile; perianth and staminodes absent; ovary erect, stigma sessile, filiform. Achene with a narrow, slightly unilateral wing.

  A genus of 2–3 species occurring in S. Africa and the mountains of eastern tropical Africa, as far north as N. Ethiopia.

Didymodoxa caffra (Thunb.) Friis & Wilmot-Dear in Bol. Soc. Brot. ser. 2, **58**: 210 (1985). —Friis in
F.T.E.A., Urticaceae: 60 (1989). TAB. **34**. Type from South Africa (Cape Province).
Urtica caffra Thunb., Prodr. Pl. Cap. **1**: 31 (1794). Type as above.
Australina acuminata Wedd. in Ann. Sci. Nat., Bot. Sér. 4, **1**: 212 (1854). —Rendle in F.T.A. **6**, 2:
306 (1917). —N.E. Br. in F.C. **5**, 2: 555 (1925). —Merxm. in Prodr. Fl. SW. Afr. 17, Urtic.: 2 (1967).
—Friis & Jellis in Kew Bull. **39**: 600 (1984). Type from S. Africa.

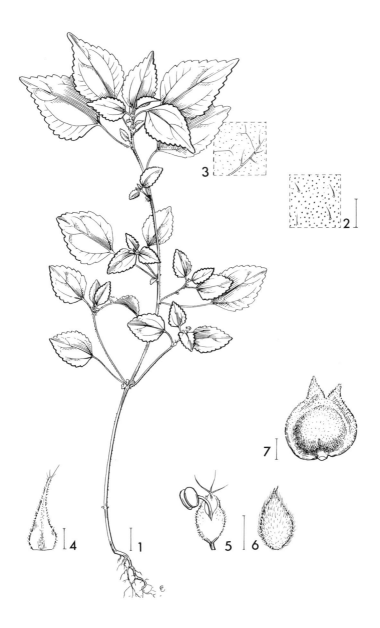

Tab. 34. DIDYMODOXA CAFFRA. 1, habit; 2, leaf, detail of upper surface; 3, leaf, detail of lower surface; 4, stipule; 5, male flower; 6, female flower, with remains of withered stigma, 1–6 *Gachoti* 678; 7, "double fruit", with 2 achenes fused along the rounded, unwinged side, *Mueller & Giess* 30244. Bar lengths: 1 = 1 cm., 2–7 = 1 mm. Drawn by Eleanor Catherine. Redrawn from F.T.E.A.

Didymodoxa acuminata (Wedd.) Wedd. in Archiv. Mus. Nat. Hist. Nat., Paris 9 (Monogr. Urtic.): 549 (1856). Type as above.

Didymodoxa cuneata Wedd. in DC., Prodr. **16**, 1: 235/62 (1869) nom. illegit., superfl., based on *Australina acuminata* Wedd.

Droguetia umbricola Engl. in Pflanzenw. Ost-Afr. **C**: 164 (1895). Type from Tanzania.

Pouzolzia erythraeae Schweinf. in Bull. Herb. Boiss. **4**, Appendix 2: 146 (1896). —Rendle in F.T.A. **6**, 2: 294 (1917). Type from Ethiopia.

Australina caffra (Thunb.) Prain in Ann. Bot. **27**: 388 (1913). Type as above.

Pouzolzia piscicelliana Buscalioni & Muschler in Engl., Bot. Jahrb. **49**: 465 (1913). —Piscicelli, Noll. Rog. Laghi. Equat.: 108 (1913). —Rendle in F.T.A. **6**, 2: 294 (1917). —Hauman in F.C.B. **1**: 215 (1948). Type: Zambia/Zaire, between Bwana Mkubwa (Buana Mukuba) and Secontwe (Sekontui), 1200 m., 30.i.1910, *Hélene d'Aosta* 512 (holotype B,†).

Erect annual herbs to c. 30(80) cm. tall, sometimes branched from the base, sometimes stems or branches rooting at the lower nodes, puberulous. Leaves alternate, lower 1–2 pairs opposite, 2–4(7.5) × 1–2.5(4) cm., ovate; apex long acuminate; base cuneate to rounded; margin crenate; lamina sparsely pilose above, and on the nerves beneath, hairs appressed. Petiole 0.8–4.7 cm. long, puberulous; stipules up to 2–5 mm. long, lanceolate. Inflorescences usually bisexual, consisting of a dense, ebracteate cluster of male flowers in the axil of one stipule and a bracteate cluster of female flowers in the axil of the other stipule; sometimes also a few flowers on a short axillary branch with much reduced leaves, or only female flowers in a cluster in the leaf axil. Bracts narrowly lanceolate, often crowned by bristles. Male perianth c. 1 mm. long, with 2–3 apical bristles. Female flower c. 2 mm. long, stigma c. 1 mm. long, sessile, filiform. Achene ovoid, brownish, c. 2 mm. long.

Zambia/Zaire. W: Zambia/Zaire border, between Bwana Mkubwa (Buana Mukuba) and Secontwe (Sekontui), 1200 m., 30.i.1910, *Hélene d'Aosta* 512 (B,†). W: between Kabwe (Broken Hill) and Bwana Mkubwa, 18.i.1910, *Piscicelli* s.n., not traced at BOLO and FT.

Also in Ethiopia, Kenya, Tanzania, Zaire and south to Namibia, and S. Africa (Transvaal, Natal, Cape Province). High rainfall woodland, in shade amongst rocks.

I have found no other specimens of this species from the Flora Zambesiaca area, and it could therefore be argued that the species should not be included in the account, especially because the publication by Buscalioni & Muschler contains species based on notorious falsifications by Muschler who took material from other herbaria, e.g. the herbarium of Schweinfurth, and published this material as if it were collected by the Duchess of Aosta's party. However, the species is badly under-collected in the major part of its range, and occurs widespread both to the north and the south of the Flora Zambesiaca area. It is therefore very likely that more material will be collected in the Flora Zambesiaca area.

159. CASUARINACEAE

By C.M. Wilmot-Dear

Trees (shrubs), dioecious less often monoecious. Branches of 2 kinds: normal persistent woody branches and deciduous green branchlets, the latter thin, flexible and little branched. Leaves on both kinds of branches reduced to whorls of small triangular scale leaves united at the base (TAB. **36**, fig. 3), becoming free on the persistent branches; midribs decurrent to the node below giving a ribbed or grooved appearance to the internode; leaf whorls, and therefore also the ribs, alternating at consecutive nodes. Flowers grouped into unisexual inflorescences with closely spaced, alternating whorls of bracts similar to the scale leaves. Male inflorescence spicate (TAB. **36**, fig. 2), solitary and terminal on deciduous branchlets (rarely also axillary on woody branches), cylindrical but tapering to a sterile basal region; flowers sessile and solitary in the axil of each bract of a whorl, a pair of lateral bracteoles enclosing each flower. Female inflorescence globose or ovoid (TAB. **36**, fig. 2), shortly stalked or subsessile, axillary toward the ends of woody branches; flowers sessile in the axil of each bract of a whorl; bracteoles as in male flowers. Male flower: a single stamen enclosed in bud by 1 or 2 (anterior and posterior) concave or cucullate, membranous perianth segments which fall as the stamen develops; mature anther exserted. Female flower: perianth 0; ovary 1-locular with a short terminal style; stigmas 2, long slender, well-exserted at maturity. Infructescence cone-like, globular, ovoid or cylindrical; ± woody due to the enlargement and thickening of accrescent bracts and bracteoles of individual flowers, the latter much the larger (often with a dorsal protuberance) and forming pairs of valves enclosing the true fruit and opening when ripe. Fruit a dark brown to black and shiny or pale grey-fawn and rather dull samara, laterally compressed, bearing a large ± translucent wing with a single longitudinal nerve excurrent at the apex.

A family of 3 genera (or 4, see note on *Allocasuarina* below), some authors recognising only one. Native to Australia, SE. Asia and Polynesia, with one species extending to Madagascar and the east coast of tropical Africa. Occurring mainly in dry, infertile and saline areas. Easily identified by the "*Equisetum*-like" green deciduous branchlets.

CASUARINA L.

Casuarina L., Amoen. Acad. **4**: 143 (1759). —Adans., Fam. Pl. **2**: 481, 543 (1763). —Forst., Char. Gen.: 103, t. 52 (1776). —Friis in Taxon **29**: 500 (1980). —L.A.S. Johnson in Fl. Australia **3**: 100–110 (1989).
Allocasuarina L.A.S. Johnson in J. Adelaide Bot. Gard. **6**: 73 (1982); in Fl. Australia **3**: 110–174 (1989) (see note below).

Trees (shrubs). Internodes of deciduous branchlets with narrow grooves. Fruit bracteoles thick or thin, with or without a dorsal protuberance; bracts relatively inconspicuous, not elaborately thickened. Samaras dark brown or black and shiny, or pale grey or fawn with surface pattern of minute, slightly raised, wavy lines.

A genus of some 70 species, distributed as for the family. Many species are widely cultivated as ornamentals and for timber and fuel wood. The nodular roots fix atmospheric nitrogen and this combined with the rapid growth of some species means that they are widely grown in many parts of the tropics. As they often hybridise in cultivation they can be difficult to name.

Allocasuarina L.A.S. Johnson
Johnson, in Journ. Adelaide Bot. Gard. **6**: 73 (1982), transfers some 58 species from *Casuarina* to the new genus *Allocasuarina;* a genus which includes all the shrubs and may be distinguished by thicker, often dorsally-thickened infructescence valves and by dark brown or black shiny seeds. For practical purposes however, a broad concept of *Casuarina* is retained here with *Allocasuarina* included in it (*C. littoralis* and *C. verticillata* would belong to *Allocasuarina*).

The species recorded from south tropical Africa, cultivated or naturalised, are included in the following key.

1. Deciduous branchlets with 14–15 ribs; scale leaves appressed, tips free for 0.5–0.7 mm.; samaras brownish, not shiny - - - - - - - - - - *Casuarina glauca* (cultivated)
- Deciduous branchlets with (6)7–10(12) ribs; scale leaves and samaras various - - 2
2. Deciduous branchlets 0.8–1 mm. in diam.; scale leaf tips free for 1–1.3 mm.; samaras dark brown, shiny - - - - - - - - - *Casuarina verticillata* (cultivated)
- Deciduous branchlets 0.4–0.6(0.9) mm. in diam.; scale leaf tips free for 0.3–0.7 mm.; samaras various - - - - - - - - - - - - - - 3
3. Infructescence valves with a large triangular transverse dorsal ridge c. 1 mm. high (TAB. **35**, fig. 4); samaras rich red-brown, shiny; whorls of male inflorescence bracts 0.5 mm. or less in diam., not crowded and hardly overlapping the adjacent whorls, axis often visible between whorls; scale leaves and male bracts uniformly pale - - *Casuarina littoralis* (cultivated)
- Infructescence valves not dorsally thickened nor transversely ridged; samaras pale, not shiny; whorls of male inflorescence bracts 1–2 mm. in diam., usually closely spaced and overlapping; scale leaves various - - - - - - - - - - - - - 4
4. Scale leaves on deciduous branchlets uniformly pale, tips free for 0.5–0.7 mm.; internodes with (6)7–8 prominent sharp ribs - - - - - - - *Casuarina equisetifolia*
- Scale leaves on deciduous branchlets with a distinct transverse brown band, tips free for 0.3–0.5(0.6) mm.; internodes with 7–9 inconspicuous rounded ribs
  - - - - - - - - - - - *Casuarina cunninghamiana* (cultivated)

Casuarina cunninghamiana Miq.

Tree. Persistent branches with 7–9 ribs; free parts of scale leaves 1–2.5 mm. long, reflexed. Deciduous branchlets 0.4–0.6 mm. in diam., with 7–9 inconspicuous ribs; scale leaves with a distinct dark red-brown band, free part 0.3–0.5(0.6) × 0.1–0.2 mm. Male inflorescence 1–1.5 mm. in diam.; bracts appressed, flowers 6–8 per whorl. Infructescences 8–12 × 7–10 mm., ovoid; valves with several irregular longitudinal fine wrinkles. Samaras pale brown, dull.

Cultivated, usually as an ornamental: **Zimbabwe**. C: Gweru Distr., Umvuma road, street tree, fr. 30.x.1966, *Biegel* 1402 (K; LISC; SRGH). **Malawi**. S: Zomba, fr. 4.iii.1981, *Chapman & Patel* 5575 (K; MAL).

Casuarina glauca Sieb. ex Spreng.

Tree. Persistent branches with c. 13 ribs; free parts of scale leaves c. 2.5 mm. long, reflexed. Deciduous branchlets 0.7–0.9 mm. in diam. with 14–15 inconspicuous ribs; scale leaves appressed, banded as in *C. cunninghamiana*, free part 0.5–0.7 × 0.15 mm. Male inflorescence c. 2 mm. in diam., whorls of bracts 10–many, flowers c. 14 per whorl; bracts ± reflexed, free for 1.6–2 mm., banded as in leaves. Infructescence 10–15 × 10–15 mm., cylindrical or depressed-ovoid; valve backs, where visible, with several fine longitudinal ridges. Samaras pale brownish, not shiny.

Cultivated: **Mozambique**. M: Maputo, fl. 27.viii.1963, *Balsinhas* 598 (K; LISC).

Casuarina littoralis Salisb. [*C. suberosa* Otto & Dietr.; *Allocasuarina littoralis* (Salisb.) L.A.S. Johnson].

Tree. Persistent branches with 7 ribs; free part of scale leaf 2.5–3 × 0.4 mm., reflexed. Deciduous branchlets 0.5 mm. in diam.; ribs 8, often prominent; scale leaves straw-coloured (in dry state), apex sometimes indistinctly darkened, free part c. 0.7 × 0.2 mm. Male inflorescence up to 0.5 mm. in diam., axis often visible between widely spaced whorls, flowers c. 7 per whorl. Infructescence 17–25 × 12–15 mm., long-cylindrical; valve backs with a large transverse ridge-like protuberance (TAB. **35**, fig. 4). Samaras rich red-brown, shiny.

Cultivated: **Mozambique**. MS: Gazaland, fl. ?date (c. 1900), *Swynnerton* s.n. (BM).

Casuarina verticillata Lam. [*C. quadrivalvis* Labill; *C. stricta* Ait; *Allocasuarina verticillata* (Lam.) L.A.S. Johnson].

Tree. Persistent branches with (7)9(12) ribs; free part of scale leaf 1.4–2 × 0.3–0.4 mm., little reflexed. Deciduous branchlets 0.8–1 mm. in diam., ribs 9–10(12) inconspicuous; scale leaves somewhat reflexed, free part 1–1.3 × 0.2 mm., rather indistinctly darkened towards the apex. Male inflorescences 2–3 mm. in diam., whorls 10–many, often little overlapping, flowers c. 10 per whorl. Infructescence 20–35 × 20 mm., ovoid-elongate; valve backs much thickened, sometimes with a small ± triangular dorsal thickening near the base, usually irregularly longitudinally wrinkled. Samaras very dark brown, shiny.

Cultivated: **Zambia**. C: Lusaka Forest Nursery, st. 5.iii.1952, *White* 2193 (K).

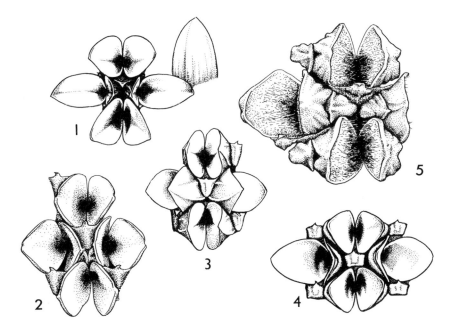

Tab. 35. CASUARINA spp. — portions of infructescences showing backs of "valves" (bracteoles). 1, C. CUNNINGHAMIANA (× 6), *Ngoundai* 84; 2, C. JUNGHUHNIANA (× 6), *Ngoundai* 254; 3, C. EQUISETIFOLIA (× 4), *Whellan* 1851; 4, C. LITTORALIS (× 4), *Morrison* 5414 (Australia); 5, C. TORULOSA (× 4), *Stauffer & Everist* 5510 (Australia). Drawn by Eleanor Catherine. From F.T.E.A.

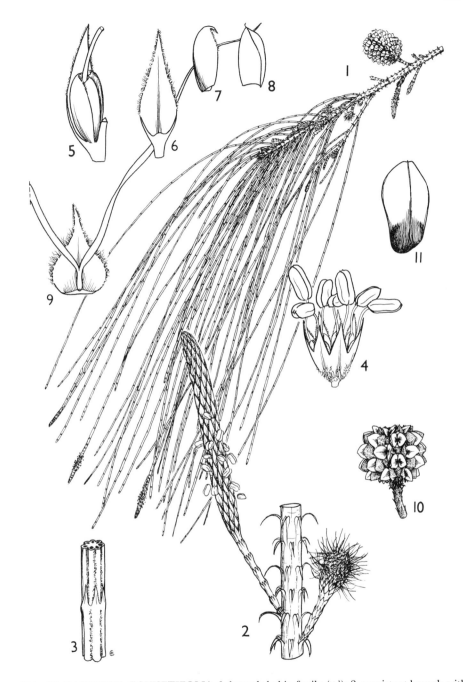

Tab. 36. CASUARINA EQUISETIFOLIA. 1, branch habit, fertile (× ½); 2, persistent branch with male and female inflorescences (× 3); 3, portion of deciduous branchlet (× 10); 4, single whorl of male inflorescence (× 10); 5, male flower (anther removed) (× 20); 6, bract (× 20); 7, bracteole (× 20); 8, perianth of male flower (× 20); 9, female flower (× 20), 1–9 *Whellan* 1851; 10, infructescence (× 1½), *Greenway* 9619; 11, samara (× 4), *Eggeling* 6628. Drawn by Eleanor Catherine. From F.T.E.A.

Casuarina equisetifolia L., Amoen. Acad. **4**: 143 (1759). —Engl., Pflanzenw. Ost-Afr., **C**: 159 (1895). —Wright in F.T.A. **6**, 2: 315 (1917). —Brenan, Check-list For. Trees Shrubs Tang. Terr.: 122 (1949). —Cufodontis, Enum. Pl. Aeth.: 2 (1953). —Dale & Greenway, Kenya Trees & Shrubs: 130, t. 26, photo 26 (1961). —Hutch., Gen. Fl. Pl. **2**: 142 (1967). —Ross, Fl. Natal: 147 (1972). —Dyer, Gen. S. Afr. Fl. Pl. **1**: 29 (1975). —Wilmot-Dear in F.T.E.A., Casuarinaceae: 5 (1985). —Gilbert & Wilmot-Dear in Fl. Ethiopia **3**: 262 (1989). TAB. **36**. Type: Rumph., Herb. Amb. 3(4), t. 57 (1743). See Friis in Taxon **29**: 499 (1980) for discussion on the nomenclature of this species.

Tree 7–25 m. tall, monoecious; stems stout, up to c. 30 cm. diam. breast height, sometimes several, bark grey-brown. Youngest persistent branches with internodes usually 1.5–2.5 mm. long; scale leaves (6)7–8 per whorl, greenish or straw-coloured, united at base into a short tube c. 1 mm. long, free part 2–3 mm. long, much reflexed, thickly chartaceous, pubescent. Deciduous branchlets numerous, especially toward the apices of the persistent branches, (0.4)0.5–0.7 mm. in diam., 7–8-ribbed; scale leaves (6)7–8 per whorl, greenish or straw-coloured, free part c. 0.5–0.7 mm. long, appressed, thinly chartaceous, glabrous, margin ciliate. Male inflorescences abundant on deciduous branchlets (rare on persistent branches), 10–30(40) × 1.2–2 mm. (width excluding exserted anthers), cylindrical, tapering gradually into a short sterile basal part; whorls of bracts 15–25; bracts c. 7 per whorl, fewer near the base, appressed, 1.1–1.8 × 0.4–0.5 mm., pubescent outside; bracteoles 0.7–1 × 0.3 mm., ovate, acute, erose-dentate-ciliate in upper one-third; perianth segments 2, up to 0.7 × 0.4 mm., rounded, membranous; filaments exserted by 1.5 mm.; anthers 0.8–1 mm. long, brownish. Female inflorescences axillary, toward the apices of younger persistent branches, scattered or densely clustered, often on same branches as male inflorescences; heads 3–5 mm. long, ovoid or subglobose; stalk 3–10 mm. long; bracts as for the male inflorescences; stigmas exserted by 3–4 mm., red. Infructescences 8–17(25) × 10–16 mm., shortly cylindrical or subglobose, ± flattened at the apex, (6)8–9(12) whorls of paired infructescence valves; valve pairs 7–8 per whorl, 1.5–3.2 mm. wide (but smaller and fewer towards apex), valves ± obovate, acute to mucronate, apices projecting by 1.5–3 mm., gap between adjacent pairs 0.5–1 mm. wide, valve backs with 2(3) longitudinal ridges (TAB. **35**, fig. 3). Samaras pale brown, dull, 5–7 mm. long, to 1 mm. thick; wing 3.5–4.5 × 2–3 mm. (those from small valves smaller), whitish or pale-brown translucent, longitudinal nerve excurrent into a mucronate curved apex.

Mozambique. N: Rovuma Bay, fr. iii.1861, *Kirk* s.n. (K); Mogincual, fr. 28.iii.1964, *Torre & Paiva* 11457 (LISC). MS: Gazaland, fl. ?date (c. 1900), *Swynnerton* s.n. (BM). M: Inhaca Isl., c. 37 km. (23 miles) E. of Maputo, fl. & fr. 5.x.1957, *Mogg* 27685 (K).

Indigenous in Malaysia, Australasia and Polynesia, cultivated widely throughout the world in tropical and warm-temperate regions to stabilise coastal dunes and as an ornamental. Indigenous or of very ancient introduction (fruits ?sea-borne) in coastal east and south-east Africa and Madagascar; has become naturalised at least in parts of coastal south-eastern Natal. Occurs in sandy areas and along the seashore.

160. SALICACEAE

By C.M. Wilmot-Dear

Dioecious trees or shrubs. Leaves alternate, deciduous; stipules small or foliaceous, usually deciduous. Flowers grouped into unisexual catkins, often appearing before the leaves; each flower subtended by a bract; perianth absent; disk present, often forming one or more fleshy glands. Male flower with 2–many stamens, filaments filiform, free or united, anthers 2-thecous, dehiscing longitudinally. Female flower: ovary 1-locular with 2–4 parietal placentas, ovules numerous, style 2–4-fid. Fruit a capsule, dehiscing longitudinally into 2–4 valves. Seeds numerous, very small, with a large tuft of long hairs arising from the funicle; embryo straight; endosperm absent.

A family comprising 2 genera and over 500 species. Widely distributed and often abundant in the N. temperate and subarctic zones, scarce and mostly confined to highland areas in the tropics; absent from Australasia and New Guinea. Chiefly of moist or wet habitats. Only one genus, *Salix*, occurs naturally in south tropical and temperate Africa, but several species of both genera are sometimes cultivated.

Buds with several clearly visible unequal outer scales; mature leaves usually relatively broad; floral bracts apically serrate or laciniate; male disk cup-shaped or annular; cultivated plants - - - - - - - - - - - - - - - - **Populus**
Buds with 1 outer scale; leaves narrow; floral bracts entire; male disk of 1–2 small distinct glands - - - - - - - - - - - - - - **Salix**

Key to the species of Populus cultivated in the Flora Zambesiaca area

1. Leaves elliptic, apex rounded to acute, base narrowly rounded to narrowly cordate, margin regularly finely bluntly serrate with incurved teeth; lamina glabrous, or pubescent only on veins - - - - - - - - - - *Populus maximowiczii* A. Henry
– Leaves either deltoid to deltoid-ovate, with base very broadly cuneate to truncate or cordate, and apex elongated, or leaves broadly ovate to orbicular with margin irregularly very coarsely serrate to bluntly sinuately-lobed or deeply palmatilobed and lamina tomentose at least when young - - - - - - - - - - - - - - - - - 2
2. Young leaves white- or grey-tomentose on one or both surfaces - - - - - 3
– Young and mature leaves glabrescent or glabrous, never tomentose on either surface - - - - - - - - - - - - - - - - - 4
3. Mature leaves (excluding those on short lateral spurs) persistently strikingly white-tomentose beneath and distinctly palmatilobed - - - - - - - *Populus alba* L.
– Mature leaves glabrescent or thinly greyish-tomentose, coarsely bluntly sinuate-dentate but rarely palmatilobed - - - - - *Populus x canescens* (Ait.) Sm. (*alba* x *tremula*)
4. Tree conspicuously columnar-fastigiate with erect or sharply ascending branches - - - - - - - - *Populus nigra* L. var. *italica* Meunch.
– Tree not fastigiate, branches spreading or downcurved - - - - - - - 5
5. Leaves deltoid to hastate, base truncate to subcordate usually with (at least in some leaves) 1–2 small glands at the junction with the petiole, margin distinctly serrate; crown of tree wide and fan-shaped with spreading or ascending branches; trunk without burrs - - - - - - - - - - - - - *Populus deltoides* Marsh.
– Leaves deltoid-ovate, base broadly cuneate or truncate without glands, margin often indistinctly serrate; crown of tree broad and rounded with massive, usually downcurved branches; trunk often with large swellings or burrs - - - - - - - *Populus nigra* L.

SALIX L.

Salix L., Sp. Pl. **2**: 1015 (1753); Gen. Pl., ed. 5: 447 (1754). —Meikle, Willows & Poplars of Great Britain & Ireland; BSBI Handbook 4 (1984).

Trees or shrubs, rarely somewhat herbaceous. Buds with one outer scale. Leaves narrow. Flowers often appearing before the leaves; disk of 1–2 fleshy glands (nectaries). Male flowers with 2–many stamens; anthers ovate or oblong. Female flowers with style divided into 2 entire or bifid stigmas.

A genus of about 400 species distributed as for the family. Fewer than 20 species occur naturally in Africa, with one native species and 2–3 cultivated species recorded for the Flora Zambesiaca region. *S. babylonica*, an introduced species, has become naturalised in parts of Zimbabwe.

Identification of *Salix* species is made difficult by the readiness of the different species to hybridise, and by the variation between the leaves produced early in the season on flowering shoots and the usually larger leaves produced in summer on sterile shoots (see Burtt Davy in Journ. Ecol. **10** (1922.)). A complete revision of the genus in Africa, especially in southern Africa, is necessary.

Stems dark red-brown or blackish; stipules minute, 0.2–0.3 × 0.4 mm., suborbicular and irregularly lobed, soon caducous; leaves narrowly elliptic to narrowly obovate, apex acute with or without a fairly abrupt acumen up to 3 mm. long; catkins 3–8 cm. long; stamens 5–8; ovary with a distinct stalk at least 0.5 mm. long, surrounded and partly enclosed by a horseshoe-shaped almost cup-like gland - - - - - - - - - - - - - 1. *subserrata*
Stems light yellowish-brown; stipules 8–12 × 1.2 mm., lanceolate, regularly serrate, persistent at least on late-season non-flowering shoots; leaves narrowly ovate-lanceolate to linear-lanceolate, narrowing gradually to a fine acumen up to 2 cm. long; catkins 2–4 mm. long; stamens 2; ovary subsessile, gland flat, adaxial, obtrapesiform 2. *babylonica* (cultivated/naturalised)

1. **Salix subserrata** Willd., Sp. Pl. **4**: 671 (1806). —Brenan, Check-list For. Trees Shrubs Tang. Terr.: 546 (1949). —F.W. Andr., Fl. Pl. Anglo-Egypt. Sudan **2**: 250 (1952). —Meikle in F.W.T.A. ed. 2, **1**: 588 (1954). —F. White, F.F.N.R.: 20 (1962). —von Breitenbach, Indig. Trees S. Afr. **2**: 70 (1965). —Friedr.-Holz. in Merxm., Prodr. Fl. SW. Afr. 14 (1967). —Palmer & Pitman, Trees of S. Afr. **1**: 413, 415 (1972). —K. Coates Palgrave, Trees Southern Africa: 92, t. 14 (1977). —Wilmot-Dear in

F.T.E.A., Salicaceae: 1 (1985). —Meikle in Fl. Ethiopia **3**: 258 (1989). TAB. **37**. Type from Egypt.
  Salix safsaf Forssk. ex Trautv., Salic.: 6, t. 2 (1836). —Thonner, Fl. Pl. Afr.: 161, t. 28 (1915).
—Skan in F.T.A. **6**, 2: 318 (1917). —Burtt Davy in J. Ecol. **10**: 71 (1922); Fl. Pl. Ferns Transvaal **2**:
432 (1932). Type as for *S. subserrata.*
  Salix octandra Sieb. ex A. Rich., Tent. Fl. Abyss. **2**: 276 (1851). Type from Ethiopia.
  Salix woodii Seemen in Engl., Bot. Jahrb. **21**, Beibl. 53: 53 (1896). —Burtt Davy in J. Ecol. **10**:
71, t. 1 fig. 4–6 (1922). —Skan in F.C. **5**, 2: 577 (1925). —von Breitenbach, Indig. Trees S. Afr. **2**:
70 (1965). Type from S. Africa (Natal).
  Salix huillensis Seemen in Engl., Bot. Jahrb. **23**, Beibl. 57: 45 (1897). —Engl., Pflanzen. Afr. **3**,
1: 7 (1915). Type from Angola.
  Salix ramiflora Seemen in Engl., Bot. Jahrb. **23**, Beibl. 57: 45 (1897). —Skan in F.T.A. **6**, 2: 318
(1917). Type from Angola.
  Salix nigritina Seemen in Engl., Bot. Jahrb. **23**, Beibl. 57: 46 (1897). —Skan in F.T.A. **6**, 2: 318
(1917). Type from Angola.
  Salix wilmsii Seemen in Engl., Bot. Jahrb. **27**, Beibl. 64: 9 (1900). —Burtt Davy in J. Ecol. **10**: 71,
t. 1, fig. 1–3 (1922). Type from S. Africa (Transvaal).
  Salix crateradenia Seemen in Engl., Bot. Jahrb. **27**, Beibl. 64: 9 (1900). —Burtt Davy in J. Ecol.
10: 69 (1922). —Skan in F.C. **5**, 2: 578 (1925). Type: Botswana (1896), *Passarge* 41 (?B†). The type
specimen cannot be traced.
  Salix hutchinsii in Bull. Misc. Inf., Kew **1917**: 235 (1917); in F.T.A. **6**, 2: 320 (1917).
—Eggeling & Dale, Indig. Trees Uganda, ed. 2: 370 (1952). —Dale & Greenway, Kenya Trees &
Shrubs: 494 (1961). Type from Kenya.
  Salix woodii var. *wilmsii* (Seemen) Skan in F.C. **5**, 2: 578 (1925).

Deciduous shrub or small tree 2–10 m. high, much-branched; branches spreading and
pendulous, rather brittle. Woody stems dark red-brown or black, shallowly longitudinally
furrowed, glabrous or more rarely with short fine light brown ± erect hairs and few
inconspicuous small pale lenticels; young stems lighter, often densely short-hairy. Leaves
alternate, 11–15 × 2–3.5 cm., narrowly- to linear-elliptic, on juvenile or late season
non-flowering growth, and 2–8(12) × 0.5–1.5 cm., linear-obovate sometimes linear-elliptic
on mature growth; apex acute sometimes acuminate, mucronate; base cuneate, margin
subentire to crenate, serrate or biserrate; lamina chartaceous, markedly discolorous,
midgreen above and pale usually glaucous beneath, glabrous or sparsely to densely hairy
on midrib and sometimes both surfaces; primary veins often reddish, rather indistinct,
7–13 pairs arching and joining near the margin, often with intermediate shorter finer
veins, reticulate venation prominulous above hardly visible beneath; petiole 3–15 mm.
long often reddish, glabrous or sparsely to densely hairy as on stem. Stipules minute,
suborbicular, 0.2–0.3 × 0.4 mm., irregularly 3-lobed, soon caducous. Catkins terminal on
short axillary branches. Male inflorescence 3–9 cm. long, axis fairly stout with abundant
long fine tangled hairs. Male flowers (20) 40 or more, spirally arranged, sessile, each
subtended by an ovate densely long-hairy greenish-yellow bract 1.5–2.5 mm. long with a
rounded or 3-toothed apex; disk divided into 2 unequal, fleshy, ± ovoid, entire or
irregularly-lobed glands, 0.7–1.1 mm. long; stamens (5)6–8, filaments 3–5.5 mm. long,
hairy mainly in the lower half, anthers rather shortly oblong, 0.2–0.5 × 0.3–0.5 mm.
Female inflorescence 1.5–8 cm. long; flowers 15–30 (or more), each subtended by a
persistent ovate or obovate acute bract 2–3.5 mm. long, pubescent as in male but more
sparsely so on upper surface; disk a minute, usually 5-lobed, fleshy ± horseshoe-shaped,
almost cup-like gland surrounding the stipe base; stipe 0.5–1.5 mm. long, glabrous; ovary
2–2.5 mm. long, ovoid; style 0.5–1 mm. long; stigmas 0.3 mm. long, shortly bilobed, ±
glabrous. Mature capsule ovoid (3)4.5–6 × 2–2.5(3) mm. with stipe 0.5–3.5 mm. long,
glabrous; splitting into 2 valves along longitudinal furrows. Seed 0.6–1.4 mm. long, ovoid,
with a large tuft of long fine white hairs ± twice the seed in length.

Botswana. N: Chobe Distr., Kasane Rapids, fr. 26.vii.1950, *Robertson & Elffers* 51 (K; PRE).
Zambia. B: Sitoti Ferry, fr. 6.viii.1952, *Codd* 7411 (BM; K; PRE; SRGH). N: E. of L. Bangweulu,
Lutikila Basin, Kanchibi R., 22.ix.1922, *Michelmore* 589 (K). W: Kitwe, 19.vii.1967, *Mutimushi* 1998 (K;
NDO). C: Lusaka Distr., c. 48 km. E. of Lusaka, Chongwe R., 15.iv.1952, *White* 2678 (FHO; K;
NDO). E: Petauke Distr., Chilongozi, Luangwa Game Reserve, Luangwa R., fr. 11.x.1960, *Richards*
13332 (K; SRGH). S: c. 1.6 km. above Victoria Falls, fr. 7.vii.1930, *Hutchinson & Gillett* 3414 (BM; K;
LISC; SRGH). **Zimbabwe**. N: Darwin Distr., Kandeya Tribal Trust Land, Machingura Dip tank, fr.
4.v.1965, *Bingham* 1467 (K; LISC; SRGH). W: Bulawayo Distr., Poole's farm, 40 km. W. of Bulawayo,
fl. 10.ix.1972, *Best* 990 (K; LISC; PRE; SRGH). C: Harare Distr., Gwebi R., on Mazowe road,
22.ii.1974, *Pope* 1120 (K; SRGH). E: Mutare Distr., Odzani R., fl. 20.vii.1957, *Chase* 6642 (COI; K;
LISC; PRE; SRGH). S: Zvishavane Distr., Runde (Lundi) R., fl. 22.iv.1971, *Chiparawasha* 387 (K;
LISC; SRGH). **Malawi**. N: Chitipa Distr., Nyika Plateau, 8 km. E. of Nganda, by tributary of Wovwe R.,
fr. 2.viii.1972, *Brummitt, Munthali & Synge* WC 137 (K). C: Ntcheu Distr., Kirk Range, Bilila Stream,

Tab. 37. SALIX SUBSERRATA. 1, branch with immature leaves (× ⅔), *Eggeling* 3543; 2, male flowering shoot (× ⅔); 3, male flower (× 6); 4, male disk (× 16); 5, male bract (× 8), 2–5 *Eggeling* 6081; 6, female flowering shoot (× ⅔); 7, female flower (× 6); 8, female disk (× 16); 9, female bract (× 8), 6–9 *Paulo* 277; 10, fruiting branch (× ⅔); 11, fruit (× 4); 12, seed (× 8), 10–12 *Kimera* 21. Drawn by Eleanor Catherine. From F.T.E.A.

Gochi Village, fl. 15.iii.1956, *Adlard* 149 (COI; K). **Mozambique**. N: Lichinga (Vila Cabral), road to L. Malawi (Niassa) fr. 10.x.1942, *Mendonça* 722 (LISC).

Widespread, from Arabia, Egypt and Ethiopia southwards to Namibia and S. Africa (Transvaal, Natal, and ?Cape (see note below)): occurs near and on the banks of streams and rivers sometimes even partially submerged, in bushland and grassland; 600–2000 m.

In S. Africa and Zimbabwe plants occur in which the leaves on the flowering growth are long and narrow — up to 10 times as long as wide and linear-elliptic rather than linear-obovate. Such plants fit the description of *S. woodii* Seemen, but in Zimbabwe they intergrade with those plants with the typical *S. subserrata* leaf form. It is therefore best to follow Coates Palgrave (1977) in considering these two species as merely different extremes of *S. subserrata*.

Salix mucronata Thunb. (syn. *S. capensis* Thunb.) which occurs throughout the Cape Province of S. Africa is very similar to this species and may yet prove to be conspecific, in which case *S. mucronata*, as the earlier name, would have priority. A detailed study, including the mapping of variation, of this species complex throughout its range of distribution is necessary to understand the relationship between the taxa.

2. **Salix babylonica** L., Sp. Pl. **2**: 1017 (1753). —Burtt Davy in J. Ecol. **10**: 81 (1922); Fl. Pl. Ferns Transvaal **2**: 431 (1932). —F. White, F.F.N.R.: 20 (1962). Type from the Orient.

A cultivated or naturalised tree to c. 10 m. tall, stems yellow-brown, branches pendulous, soon glabrescent. Leaves 6–16 × 1–1.5 cm., narrowly ovate-lanceolate to linear-lanceolate, tapering gradually into a long fine acumen, base cuneate, margins regularly serrulate, lamina pale usually glaucous beneath, glabrous, lateral veins c. 15, distinct; petiole to 6 mm. long, soon glabrous; stipules 8–12 × 1–2 mm., ovate-lanceolate, serrate, persistent at least on late-season sterile shoots. Catkins up to 4 cm. long, on short axillary branches, bearing 1–few leaves at the base; the axis and the basal parts of bracts hairy. Ovary c. 1.2 mm. long, sessile or subsessile; disk gland adaxial, flat, obtrapesiform and ± undulate at the apex; style short, stigmas 2-lobed. Mature capsule 2–3 mm. long.

Cultivated specimens: **Botswana**. SE: 4a Ngwaketse, Kanye, fr. 20.xii.1972, *Kelaole* A.104 (SRGH). **Zimbabwe**. C: Harare, cultivated in Greendale, fl. 20.ix.1969, *Dell* in GHS 2000867 (K; PRE). E: Penhalonga, Strickland's Farm, fl. & fr. 14.ix.1934, *Gilliland* K.789 (BM; K) (maybe a hybrid with *S. subserrata*).

This species was introduced from Europe and is planted extensively in southern Africa along rivers and around dams. It (and possibly its hybrids) has become naturalised in parts of Zimbabwe and S. Africa.

Biegel 4421 (K; SRGH) from Goromonzi in Zimbabwe, is a sterile specimen from a cultivated shrub and is probably *S. medenii* Boiss. — a species distinguished by its large, broad, elliptic leaves, 3–5 × 2–2.5 cm. with dense, erect pubescence beneath.

161. CERATOPHYLLACEAE

By C.M. Wilmot-Dear

Submerged aquatic, branching, usually rootless herbs, perennating by buds, monoecious. Leaves in whorls, filiform, once or more dichotomously branched, margins ± spinose-dentate; apical segments truncate, 2-spined at the apex with a reddish glandular projection between the spines (basal segments sometimes parasitised, becoming swollen and sac-like). Flowers axillary, 1–several per node, male and female flowers at different nodes, ± sessile; perianth lobes 6–13, united at the base, strap-shaped or obovate, margin often with a single lateral hyaline spine or ± lacerate, apex as for leaf apex. Male flowers subsessile; stamens up to c. 30 in several whorls on a domed torus around the pistillode; filaments short or absent; anthers oblong, extrorse, 2-locular, dehiscing longitudinally, connective produced apically into 2 spines and a central projection as in perianth lobes, immature anthers resembling perianth lobes, margins 1–3 spined; mature anthers detach and float to the water-surface where they dehisce and shed pollen on to plants below. Female flowers sessile or shortly pedicellate; staminodes absent; ovary superior, sessile, ovoid, tapering into a long style; ovule 1, pendulous. Fruit a nut, ovoid or ellipsoid, often warty and basally-spined, style ± persistent, spinose. Embryo straight, endosperm absent.

A family world-wide in its distribution, comprising one genus only.

CERATOPHYLLUM L.

Ceratophyllum L., Sp. Pl. **2**: 992 (1753); Gen. Pl., ed. 5: 428 (1754). —Wilmot-Dear in Kew Bull. **40**: 243–271 (1985).

Characters as for the family.

A genus of two species worldwide. Other authors have recognised between 1 and 10 (or more) species, but most have accepted 3, of which *C. muricatum* Cham. now seems best reduced to a subspecies of *C. submersum*. At least one variety of each species occurs in the Flora Zambesiaca region (but see note under *Ceratophyllum submersum* L. below).

Leaves branching twice, rarely once or (in lower parts, especially of main axis) thrice, spiny teeth on leaf margins often many, prominent; mature fruit with a long apical spine and 2 prominent basal spines, lateral flattening of fruit slight, surface ± smooth - - - *1. demersum*
Leaves, at least the majority on all parts of the plant, 3–4-times branched, spiny teeth on leaf margin few, always small, inconspicuous; mature fruit with a long or very short apical spine, and with or without basal and marginal spines, lateral flattening of fruit ± well-marked, forming a longitudinal marginal rim or wing, the surface strongly papillose or warty *2. submersum*

1. **Ceratophyllum demersum** L., Sp. Pl. **2**: 992 (1753). —Engl. in Mildbr., Wiss. Ergebn. Deutsch. Zentr.-Afr. Exped. **2**: 206 (1911). —Medley Wood, Natal Pl. **6**, plate 551, cum tab. (1912). —Skan in F.T.A. **6**, 2: 326 (1917); in F.C. **5**, 2: 580 (1925). —Robyns, Fl. Sperm. Parc Nat. Alb. **1**: 167 (1948). —Andrews, Fl. Pl. Anglo-Egypt. Sudan **1**: 14, fig. 13 (1950). —Hauman in F.C.B. **2**: 165 (1951). —Cufodontis in Enum. Pl. Aeth.: 106 (1953). —Keay in F.W.T.A. ed. 2, **1**: 65 (1954). —Friedr.-Holzhammer in Merxm. Prodr. Fl. SW. Afr., fam. 40 (1968). —Lind & Tallantire, Common Fl. Pl. Uganda: 114 (1971). —Agnew, Upl. Kenya Wild Fl.: 81 (1974). —Gibbs Russell in Kirkia **10**: 472 (1977). —Raynal-Roques in J.R. Durand & Leveque, Fl. Faun. Aquat. Sah.-Soud.: 103 (1980). —Wilmot-Dear in F.T.E.A., Ceratophyllaceae: 3 (1985). Type from Europe.

Aquatic herb to c. 3 m. long; main stem to 2 mm. in diam., robust and wiry, rarely delicate, finely longitudinally grooved. Leaves bright or olive green, 7–11 per whorl, 8–40 mm. long, branching dichotomously once or twice, (sometimes 3 times in lower part of plant), 0.2–0.7 mm. (1–2 cells) thick, lowest segments sometimes to 1 mm. thick, the apical and often the lower segments with many, rarely few, spine-tipped marginal teeth (0.1)0.2–0.5 mm. long. Male flowers 1–3 per node, often many per branch, variable in size, up to 2.5(3.5) mm. in diam.; perianth broadly cup-shaped, lobes 0.5–1.3 × 0.2–0.4 mm. with a glandular projection to 0.2 mm. long; stamens up to c. 30, anthers subsessile, 1–2 × 0.4–1.5 mm. when mature; pistillode c. 0.6 mm. long. Female flowers (? always) 1 per whorl, few per branch; perianth closely surrounding the ovary, persistent in fruit, lobes resembling those of male perianth, glandular projection to 0.7 mm. long. Ovary to 1 × 0.6 mm., smooth; style usually over 2 mm. long. Fruit ovoid or ellipsoid, rarely obovoid, with an apical spine (1.5)3.5–9 mm. long, usually with 2 prominent basal spines, surface smooth with sparse or numerous slightly raised dark dots.

Var. **demersum** TAB. **38**, figs. A1–11.
Ceratophyllum oxyacanthum Cham. in Linnaea **4**: 504, t. 5 fig. 6b (1829) non Schur. Type from Germany.
Ceratophyllum demersum var. *oxyacanthum* (Cham.) K. Schum. in Mart., Fl. Bras. **3**, 3: 748 (1894). —Engl., Pflanzenw. Ost-Afr. **C**: 178 (1895). Type as for *Ceratophyllum oxyacanthum*.
Ceratophyllum tuberculatum Cham. in Linnaea **4**: 504, t. 5 fig. 6d (1829). Type from India.

Fruit very slightly laterally flattened and without a marginal rim, 4–5.5 × 3–3.5 mm.; apical spine (1.5)3.5–9 mm. long; basal spines present, (0.5)1.5–6 mm. long.

Caprivi Strip. Katima Mulilo, Liambezi Meer, st. vii.1974, *Pienaar & Vahrmeijer* 473 (K; PRE). **Botswana**. N: Matlapaneng, 8 km. NE. of Maun, st. 19.iii.1965, *Wild & Drummond* 7163 (K; LISC; PRE; SRGH). SE: Mopipi Flats, fr. iv.1977, *Allen* 416 (PRE). **Zambia**. N: Mansa Distr., L. Bangweulu, Luaba Lagoon, st. 3.ii.1959, *Watmough* 221 (K; LISC; PRE; SRGH). W: Kitwe Distr., Mwekera Dam, st. 17.viii.1962, *Mortimer* 194 (SRGH). C: Luangwa South Game Reserve, Mfuwe Lagoon, Luangwa R., st. 27.iv.1965, *B.L. Mitchell* 2701 (K; SRGH). E: Msuzi R., Msuzi Weir, st. 29.iv.1961, *Mortimer* 180 (K; SRGH). S: Gwembe, st. 3.i.1965, *Edwards* in GHS 159924 (LISC; PRE; SRGH). **Zimbabwe**. N: Binga Distr., Sinamwenda R., Elephant Isl., st. 27.ix.1970, *D.S. Mitchell* 1270 (K; PRE; SRGH). W: Hwange Distr., Katsatetsi R., Kazuma Range, st. 10.v.1972, *Gibbs Russell* 1954 (K; SRGH). C: Harare Distr., pool by Muzururu R., st. 8.i.1973, *Gibbs Russell* 2510 (K; PRE; SRGH). **Malawi**. N: Mzimba Distr., L. Kazuni, 24 km. from Mperembe, st. 21.viii.1966, *Pawek* 5 (SRGH). C: Salima Distr., Linthipe R. opposite Maleri Isl., st. 1.viii.1951, *Chase* 3852 (BM; SRGH). S: Nsanje Distr., Chiromo, st. 22.iii.1960,

Phipps 2603 (K; SRGH). **Mozambique**. N: Amaramba L., st. 5.x.1958, *Monteiro* 124 (LISC). MS: Buzi R., st. 20.vi.1961, *Leach & Wild* 11119 (K; SRGH). GI: Guijá Distr., st. vii.1915, *van Dam* s.n. (SRGH). M: Inhaca Isl., st. 31.iii.1967., *Mogg* 33610 (LISC; ?SRGH).

Also recorded from W. Africa to the Sudan and Ethiopia and south through east tropical Africa to Angola and S. Africa. Almost worldwide in distribution, but not recorded from the Malay Peninsula; introduced into New Zealand and Mascarene Islands. Occurs in static to fast-flowing, shallow or deep water, in lakes, rivers, streams and reed swamps; tolerant of brackish estuarine conditions of high salinity but not found in seasonal, highly alkaline pools; commonly associated with *Azolla*, *Pistia* and *Nymphaea*, often locally dominant, forming thick mats just below water surface; 3–1600 m.

Var. *apiculatum* (Cham.) Ascherson is recognised by the basal spines of the fruit which are reduced to minute tubercles up to 0.5 mm. long, and is recorded from Zaire, northern Europe, India, North and South America. Var. *inerme* Radcliffe-Smith, with basal spines of fruit absent, has been found in Ghana, Iraq and northern Europe. Var. *platyacanthum* (Cham.) Wimmer, whose markedly flattened fruit both possesses a distinct marginal rim running from apex to base and extending into much flattened basal spines and also has a large protuberance or flattened spine on each face, has been found in Europe and the Far East.

2. **Ceratophyllum submersum** L., Sp. Pl. ed. 2: 1409 (1763). —Hauman in F.C.B. **2**: 165 (1951). —Wilmot-Dear in F.T.E.A., Ceratophyllaceae: 4 (1985). Type from Europe.

Aquatic herb similar to *Ceratophyllum demersum*, but usually more delicate, differing as follows: Leaves usually lighter green, 2–4 cm. long, branching dichotomously 3–4 times (rarely twice in some whorls), usually rather fine, 0.1–0.3 mm. thick, to 2.5 mm. thick in lower segments, marginal teeth few, inconspicuous, often absent from lower segments, up to 0.1–0.2 mm. long. Male flowers to 2 mm. in diam.; anthers 0.6–0.8 × 0.4–0.7 mm., often smaller and relatively broader than in *C. demersum*. Female perianth lobes (1.5)1.8–2 × 0.1–0.3 mm., relatively narrower and usually longer than in *C. demersum*. Fruit usually markedly laterally flattened with a marginal rim or wing.

Botswana. N: Thaoge R. (Thauge R.), 19°48'S, 22°21'E, fl., 22.viii.1975, *P.A. Smith* 1443 (K; SRGH). **Malawi**. S: Lower Shire R., st. ii.1888, *Scott* s.n. (K). **Mozambique**. M: Inhaca Isl., fr. 23.ix.1957, *Mogg* 27491 (K).

Occurs mainly in slow-flowing or stagnant water, often in seasonal highly alkaline lakes, swamps and pools; not found in estuarine brackish conditions.

Recognition of the infraspecific taxa depends largely on characters of the mature fruit, and since the material cited above (the only material of this species from the Flora Zambesiaca area seen so far) has immature fruit, or lacks fruit, it has not been possible to determine the subspecies or variety to which the specimens belong. The taxa most likely to occur in the Flora Zambesiaca area are described below. For a full treatment and synonymy see Wilmot-Dear's revision of the genus in Kew Bull. **40**: 243–271 (1985).

Subsp. **submersum**
Var. **submersum**   TAB. **38**, figs. B1, B2.

Fruit markedly laterally flattened with a ± distinct longitudinal "marginal" rim; 3–5 × 2.5–3 mm., ellipsoid; apical spine up to 1(2) mm. long; basal spines absent; surface with numerous minute warty papillae giving a prickly appearance especially along marginal rim.

Recorded from Kenya, Tanzania, parts of Central and West Africa and possibly Zaire; also from Dominica and parts of Europe and Asia.
Var. *squamosum* Wilmot-Dear, found in E. China, differs in having large scaly warts on its fruit surface, while var. *haynaldianum* (Borbas) Wilmot-Dear occurs in Hungary.

Subsp. **muricatum** (Cham.) Wilmot-Dear in Kew Bull. **40**: 266 (1985).
Var. **echinatum** (A. Gray) Wilmot-Dear in Kew Bull. **40**: 266 (1985). Type from N. America.
      Ceratophyllum muricatum Cham. in Linnaea **4**: 504, t. 5, fig. 6c (1829). Type from Egypt.
      Ceratophyllum cristatum Perr. & Guill., Fl. Seneg. Tent.: 296 (1833). Type from Senegal.
      Ceratophyllum echinatum A. Gray in Ann. Lyceum Nat. Hist. N.Y. **4**: 49 (1837). Type as above.
      Ceratophyllum demersum var. *echinatum* (A. Gray) A. Gray, Man. ed. 2: 383 (1856). Type as above.
      Ceratophyllum demersum var. *muricatum* (Cham.) K. Schum. in Martius, Fl. Bras. **3**: 750 (1894). Type as above.

Fruit markedly laterally flattened, marginal rim extended into a ±entire or irregularly crenate to long-spinose wing, 3–4.5 × 2–3(3.5) mm., ellipsoid; apical spine (1)4–9 mm. long, 2 basal spines (0.5)2–6 mm. long; surface usually with few–many small ± elongate,

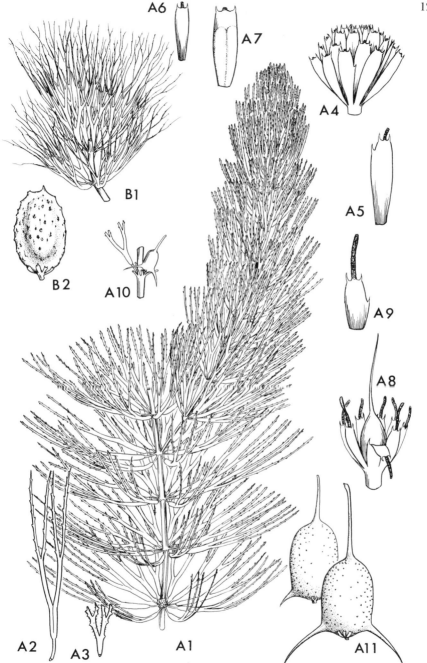

Tab. 38. A.—CERATOPHYLLUM DEMERSUM var. DEMERSUM. A1, habit (×1½), *Chandler* 2082; A2, leaf (×1½), *Milne-Redhead & Taylor* 7241; A3, leaf (×1½), *Vanderplanks* s.n.; A4, male flower (× 12); A5, male perianth lobe (× 18); A6, immature anther (× 18); A7, nearly mature anther (× 18); A8, female flower (× 12); A9, female perianth lobe (× 18), A4–9 *Bogdan* 2333; A10, fruit on stem (× 1); A11, fruits (× 4), A10–11 *Richards* 20189. B. —CERATOPHYLLUM SUBMERSUM var. SUBMERSUM. B1, portion of stem with leaf-whorls (× 1½), *Norman* 156; B2, fruit (× 4), *Richardson* 14. Drawn by Eleanor Catherine. From F.T.E.A.

rarely rounded, warty papillae up to 0.2 mm. long, sometimes spinulose and up to 0.3 mm. high.

Recorded from S. Africa (Eastern Transvaal) and Angola; and also from Senegal, Ghana, Chad, Sudan, Ethiopia, Central and North America, northern South America, Western Russia, India and Pacific Islands.

This is the typical variety of subsp. *muricatum,* but by the rule of priority has to take the existing varietal epithet *echinatum.*

Var. *manschuricum* Miki, found in Europe and the Far East, differs in having few unequal-sized large dark warts or spines on its fruit surface.

INDEX TO BOTANICAL NAMES